AF811609

The Materials Research Society Series

The Materials Research Society Series covers the multidisciplinary field of materials research and technology, publishing across chemistry, physics, biology, and engineering. The Series focuses on premium textbooks, professional books, monographs, references, and other works that serve the broad materials science and engineering community worldwide. Connecting the principles of structure, properties, processing, and performance and employing tools of characterization, computation, and fabrication the Series addresses established, novel, and emerging topics.

Christoph Klemmt · Gregory Beaucage
Editors

Biodegradable 3D-Printing

Springer

Editors
Christoph Klemmt
School of Architecture and Interior Design
University of Cincinnati
Cincinnati, OH, USA

Gregory Beaucage
Department of Chemical and Materials Engineering
University of Cincinnati
Cincinnati, OH, USA

ISSN 2730-7360 ISSN 2730-7379 (electronic)
The Materials Research Society Series
ISBN 978-981-95-4024-2 ISBN 978-981-95-4025-9 (eBook)
https://doi.org/10.1007/978-981-95-4025-9

© The Editor(s) (if applicable) and The Author(s), under exclusive license to Springer Nature Singapore Pte Ltd. 2026

This work is subject to copyright. All rights are solely and exclusively licensed by the Publisher, whether the whole or part of the material is concerned, specifically the rights of translation, reprinting, reuse of illustrations, recitation, broadcasting, reproduction on microfilms or in any other physical way, and transmission or information storage and retrieval, electronic adaptation, computer software, or by similar or dissimilar methodology now known or hereafter developed.
The use of general descriptive names, registered names, trademarks, service marks, etc. in this publication does not imply, even in the absence of a specific statement, that such names are exempt from the relevant protective laws and regulations and therefore free for general use.
The publisher, the authors and the editors are safe to assume that the advice and information in this book are believed to be true and accurate at the date of publication. Neither the publisher nor the authors or the editors give a warranty, expressed or implied, with respect to the material contained herein or for any errors or omissions that may have been made. The publisher remains neutral with regard to jurisdictional claims in published maps and institutional affiliations.

This Springer imprint is published by the registered company Springer Nature Singapore Pte Ltd.
The registered company address is: 152 Beach Road, #21-01/04 Gateway East, Singapore 189721, Singapore

If disposing of this product, please recycle the paper.

Preface

This book grew out of a collaborative project between Klemmt and Beaucage on the development of wood-based 3D printing techniques for architectural applications. During this work it became clear that the scientific, technical, and design fields were significantly overlapping in a wide range of applications centered around biodegradable materials and advanced manufacturing. In the creative disciplines including architecture, product or fashion design, but as well in fine art or medicine, materiality and fabrication play an important role for both design and research. Specifically the perceived need to increase the sustainability of products or the built environment has led to a greater interest in material science and manufacturing. As it is not only a designer's task to shape a product, but also to devise its materials and mode of production, it has become the designer's responsibility to be proficient in the possibilities of sustainable materials and processes, and to develop novel approaches in fabrication. This has led various researchers from the design disciplines to investigate sustainable production methods, including methods of 3D printing with biodegradable materials, and in the process seek the collaboration with material scientists.

At the same time, the advances made in the science fields regarding novel materials and manufacturing methods require applications to become useful for society and relevant for commercialization. Hence, engineers have an interest in working with those who explore the design possibilities of objects ranging from medical implants to buildings, leading to mutually beneficial collaborations. Even more so in areas of bioprinting, different disciplines such as manufacturing, material science, biology, and medicine become closely intertwined, with interdisciplinary collaborations forming the basis of any project.

As the different fields that produce research in this area are spread wide, yet their endeavors are closely related, we took on a project to present some of the most important and interesting work in the area of 3D printing with biodegradable materials in a book that spans disciplines and scales. We approached a broad range of contributors and were met with great enthusiasm and encouragement. The project is largely self-supported with some funding and time support from the University of Cincinnati. The book is a collection of primary examples of the application of biomaterials in 3D printing. It could be used as the basis of a course or short course on this subject. We find the advances outlined in the book to be fascinating and exciting and we hope that a general audience or those interested in the future of 3D printing will find the book of interest. The book includes an appendix with DIY instructions on how to implement some of the novel technologies discussed in the chapters. This would be of use to hobbyists or as a basis for a hands-on component to a course. We hope that you find this book interesting and that it encourages access to this expanding field.

Cincinnati, OH, USA

Christoph Klemmt
Gregory Beaucage

Contents

Part I
Introduction

Chapter 1
Intention of this Book

Christoph Klemmt and Gregory Beaucage

Abstract This chapter lays out the intention of the book. It states several of the opportunities of biodegradable materials, especially the aspect of sustainability, which is a concern of many other material options, especially thermoplastics. It then proposes possibilities and opportunities of 3D printing, a manufacturing methodology that still offers exciting areas of progress and development. Based on those, it sets out the potential benefits of 3D printing with biodegradable materials. The chapter then gives a brief overview of the individual chapters, grouped by the Introduction, Materials and Methods, and Applications. It then concludes with an overview of the ongoing and future work in the field of 3D printing with biodegradable materials.

Keywords 3D printing · Biodegradable · Materials · Additive manufacturing · Sustainability

1 Biodegradable 3D Printing

The world around us is saturated with plastics: The computer or phone you likely are reading this book on, the toys, hand cream tube, snack packaging and pens on your desk, the table and chair you are sitting on, furniture, lighting and storage in the room, the power cables, drainage pipes, window frames and cladding panels of the building you are in, endless amounts of components in your car, your fridge, at the hospital and in the urban environment. Many items we touch daily, while even more are hidden from our view, or support the ecosystems that produce the energy, products and services that feed our daily needs. Plastics are clearly of a truly

C. Klemmt (✉)
College of Design, Architecture, Art and Planning, University of Cincinnati, Cincinnati, OH, USA
e-mail: christoph@orproject.com

G. Beaucage
Chemical and Materials Engineering, University of Cincinnati, Cincinnati, OH, USA

© The Author(s), under exclusive license to Springer Nature Singapore Pte Ltd. 2026

C. Klemmt and G. Beaucage (eds.), *Biodegradable 3D-Printing*, The Materials Research Society Series, https://doi.org/10.1007/978-981-95-4025-9_1

ubiquitous nature due to their undefeated benefits and affordability. And they have gained such an enormous importance in less than 100 years.

This short time frame is testament to the benefits of plastics. Yet also, it hints at the likelihood of negative consequences of their overuse that we are still in the process of discovering. Plastic waste may be the most important aspect, and the one that has been discussed for the longest time. Even with recycling available for many polymers, the percentage of material being recycled remains small, and the amount of waste accumulating in landfills or the natural environment keeps rising. Yet also many other concerns have been raised: The limited resources of petrochemicals, polymers leaching harmful compounds, exposure to plasticizers, the world's exposure to micro and nanoplastics. It appears that new and unforeseen challenges, and the severity of their impact, are discovered every few years. We therefore face the challenge of either addressing those concerns by adjusting the polymers we use, or by finding alternative materials without those problems.

The ubiquity of polymers today makes it hard to imagine a world without any plastics, yet many people alive today were born into a world before their invention or widespread use. So, what were products and objects made from a hundred years ago? Certainly, many types of objects that we use today did not exist back then, such as electronics or medical implants, but also many items that we mainly know as plastic objects today existed in other forms. Metals, minerals including silicates, and animal products were used to produce many items, including glass and enamelware or brick buildings. Yet especially plant fibers made up a large amount of the raw materials in use, including for objects of a limited lifespan such as packaging or consumables. Wood, wicker, bamboo, card, paper, cotton, linen, rubber, oils, dyes and medicines are all commonly made from plants. This makes them renewable, mostly with a nontoxicity to humans and the environment, and biodegradable, resulting in a circularity of manufacturing without any waste. At the onset of polymer development, also the first plastics were largely made from plant materials and fall into the group of bioplastics, such as Parkesine or cellophane, though cellulose nitrate (Parkesine) and cellulose acetate are not biodegradable and cellophane with coatings or additives may be rendered non-biodegradable.

While a world made from plants instead of petrochemical materials would address many of the concerns that the plastic industry faces today, it rarely seems appropriate to step back in technological development, and going back to the past seems unlikely in addition to undesirable. Yet learning from the past has always been of benefit, including in the development of novel technologies. We therefore aim to promote the development of new methods of manufacturing with biodegradable plant materials, with this book focusing on 3D printing. This can be expanded to other biosource materials such as fungi, algae, bacteria, or chitin and its derivative chitosan which are biodegradable.

3D printing, a subgroup of but often used synonymously with Additive Manufacturing (AM), is a method of producing objects layer by layer. Material is extruded, molten, or otherwise solidified in slices until the item is completed. A wide variety of materials are used in 3D printing, including polymers, bioplastics, metals, concrete and clay. In scale, applications range from small-scale medical implants to complete building structures. While the technology used to be referred to as rapid prototyping, it is nowadays used by many companies for full-scale production. While 3D printing is not expected to fully replace other manufacturing processes, a lot of research is conducted to improve and develop the available methods and materials, and to find new areas of application. As we write this introduction, my daughter is watching an influencer's video on commercially available, AI-designed, 3D-printed shoes. Recent research provides ever better methods, precision, scale, range of novel materials and applications.

However, while 3D printing has been touted as beneficial for the environment due to only using the material that makes up the final product unlike subtractive manufacturing, the 3D printing industry faces many concerns regarding its sustainability: The use of petrochemical polymers is widespread, many methods are highly energy intensive, be it in the sourcing of the raw materials or the printing process such as the melting of metals, and the issue of waste is rarely addressed. Process waste includes raw materials or support structures, but also failed prints and cheap throwaway objects contribute to this problem. While materials such as polylactic acid (PLA) are marketed as green and compostable, it is only compostable in specialist industrial facilities when sorted free of additives and absent blends with non-biodegradable polymers, and virtually all of its waste ends up non-composing in landfills or the natural environment, where it turns into harmful micro or nano-bioplastics.

Various scientists have therefore started to investigate the possibilities of 3D printing with biodegradable materials, especially those sourced from plants or the natural environment. This results in benefits including renewable resources, circularity without any waste, as well as, in many cases, manufacturing technologies that consume little energy. However, many challenges remain, including in areas such as processability, curing, and durability of the printed objects. At the same time, the use of novel materials has opened up entirely new fields of application where traditional 3D printing materials are of no benefit or are counterproductive, such as the printing of one-off molds for casting, bioprinting that supports live organisms, or in the medical sector, where biodegradable materials offer nontoxicity or a desired natural breakdown inside of the human body.

This book therefore has the dual aim to bring together current academic research into AM with biodegradable materials, and on a social level to promote more sustainable practices and materials in 3D-printing. The research presented in this book, originating from different disciplines, has investigated 3D-printing with plant fiber, living organisms and other biodegradable materials, and the work is applied across various fields ranging from small scale medical implants to large architectural applications.

2 Outline of the Book

This book is a collection of original research chapters by the leading researchers in the field, with an introduction and conclusion by the editors. It consists of three sections: The introduction consists of three chapters. Beyond this chapter, the second is a comparison of lifecycle analyses of common 3D-printing methods and materials, and the third chapter a review of existing research. The second section focuses on methodologies and the technical means by which a 3D-printing of biodegradable materials can be achieved. Those papers include research into different fibers, binders, and technologies. A focus of this section is on material science and chemical engineering. The third section concentrates on applications in different fields, and the possibilities that the biodegradable materials can offer. This includes applications as medical implants, product design, molds for casting, and applications at an architectural scale.

2.1 Part 1: Introduction

Chapter 1: Intention of This Book (this chapter)

(Christoph Klemmt, Gregory Beaucage)

This chapter gives an introduction to the intention of this book.

Chapter 2: Sustainability in Additive Manufacturing: A Comparison of 3D Printing Methods and Materials

(Rajat Sodhi, Shuo Qin)

This chapter compares Life Cycle Assessments of different 3D printing methodologies and materials from different sources. The data give an insight into the challenges and opportunities faced by 3D printing technologies (Chap. 2).

Chapter 3: Review of Biodegradable Materials in Additive Manufacturing

(Gregory Beaucage, Christoph Klemmt)

This chapter gives an overview of available methods, materials, and experimental and ongoing research into 3D printing. It explains some of the underlying principles, as well as opportunities and research areas related to 3D printing, such as control over fiber alignment or functional grading (Chap. 3).

2.2 Part 2: Materials and Methods

This Part includes research chapters that cover some of the most important novel materials and methods that exist in the area of biodegradable 3D printing.

Chapter 4: FLAM 3D Printing with Chitin–Cellulose Composites

(Stylianos Dritsas and Javier G. Fernandez)

This chapter explains a method using cellulose and chitin composites for 3D printing. The method has been used for the production of several prototypes (Chap. 4).

Chapter 5: 3D Bioprinting with Cellulose-derived Bioink and Precursors

(Bashir Khoda and Md. Ahasan Habib)

This chapter covers the 3D printing of nano cellulose and explains how the combination of design, material and manufacturing results in successful applications for the viability of life cell cultures (Chap. 5).

Chapter 6: Living Composites: Mycelium Biofabrication and the Evolution of Design

(David Andréen)

This chapter covers the 3D printing of mycelium, whereby the fungus acts as the binder for a biodegradable substrate. The method has been used for the construction of a large-scale column (Chap. 6).

Chapter 7: Three-dimensional Printing of Lignin

(Md Shajedul Hoque Thakur and Muhammad M. Rahman)

This chapter reviews various methods of using lignin in 3D printing. The authors present their method of utilizing lignin as a binder for cellulose (Chap. 7).

2.3 Part 3: Applications

This Part covers some of the novel applications, across various scales and disciplines, that the 3D printing of biodegradable materials can be utilized for.

Chapter 8: Three-Dimensional Printing of Biodegradable Materials in the Biomedical Sector

(Karthik Tappa)

This chapter presents various examples of applications in the biomedical field. Those include naturally degrading bone scaffolds, tissue scaffolds that support life cell cultures, or biodegradation as a means for a controlled drug delivery (Chap. 8).

Chapter 9: Wet-Mold Concrete Casting with Biodegradable Materials

(Brendan Kiser, Ryan Kennedy, Lucas Palicki)

This chapter presents a method for the 3D printing of molds for concrete casting. The biodegradable printing material remains in its wet, uncured stage during the casting process (Chap. 9).

Chapter 10: 3D Printing of Wood and Bioplastic Composites for Architectural Applications

(Christoph Klemmt, Gregory Beaucage)

This chapter presents research into the large-scale 3D printing of wood. Using wood shavings and a cellulose-derived binder, it was used to produce a large-scale architectural column (Chap. 10).

Chapter 11: 3D Printing With Living Matter

(Frank Melendez and Nancy Diniz)

This chapter presents a series of art and product design projects that involve the 3D printing of live organisms. It discusses historical and theoretical implications of involving living matter in design (Chap. 11).

2.4 Part 4: Conclusions

Chapter 12: Conclusions

(Christoph Klemmt and Gregory Beaucage)

The conclusion summarized the contributions of the different chapters and highlights future trajectories of the field (Chap. 12).

2.5 Appendix: Material Compositions, Recipes, and Supplemental Information of Some of the Papers From This Book

The appendix contains information for a practical application of some of the methods presented in this book. The aim of the appendix is to enable and encourage researchers, students and hobbyists to use and further develop this work (Appendix).

3 Overview

As the field of biodegradable 3D printing is actively developing, there are many research projects that we were unable to include in this book. Commercial companies are starting to offer fully biodegradable products manufactured by means of 3D printing. At the same time, the maker community is becoming aware of waste concerns in 3D printing and is in search of sustainable alternatives. As there are still many challenges that need to be addressed, and there are many opportunities for the development of new methods, materials, and applications, this book offers an in-depth insight into the topic. Due to the concerns with current manufacturing processes and materials, as well as the opportunities it offers, we believe that biodegradable 3D printing has the potential to become a truly transformative industry.

Christoph Klemmt graduated from the Architectural Association in London and received his Doctor Technicae from the University of Applied Arts Vienna. He has worked amongst others for Zaha Hadid Architects, where his responsibilities were focused on the company's projects in China. He is Associate Professor at the Department of Design, Art, Architecture and Planning at the University of Cincinnati, where he set up and runs the Architectural Robotics Lab. He co-founded Orproject, an architect's office specializing in advanced geometries with an ecologic agenda. Orproject has been featured widely and has exhibited at the Palais De Tokyo in Paris, the China National Museum in Beijing and the Biennale in Venice.

Gregory Beaucage is Professor of Chemical and Materials Engineering at the University of Cincinnati. He has published extensively in polymers and materials science with 189 publications and 6,900 citations. He has collaborated with Christoph Klemmt on the development of new materials for 3D printing of wood for the past four years. Beaucage holds a Ph.D. from the University of Massachusetts at Amherst in Polymer Science and Engineering and was a post doc at Sandia National Laboratories before joining the faculty at the University of Cincinnati. He was a Fulbright Global Scholar at Sheffield University in the UK and at several Universities in Ethiopia, South Africa, and Lesotho. He has been Visiting Professor at University of the Witwatersrand, Johannesburg, University of Sheffield, University of Cape Town, and ETH in Zürich.

Chapter 2
Sustainability in Additive Manufacturing: A Comparison of 3D Printing Methods and Materials

Rajat Sodhi and Shuo Qin

Abstract This chapter evaluates different materials used in 3D printing, starting from its early days of stereolithography to current practices that focus on sustainability. It examines the shift from traditional plastics to eco-friendly alternatives, including ABS, PLA, bioplastics, metals, concrete and natural materials like clay, while addressing their environmental impacts and the industry's efforts to make 3D printing sustainable. Life cycle assessment of materials is used to compare each stage of the product process from material extraction to disposal to evaluate its environmental footprint. Energy consumption and environmental implications highlight challenges such as energy-intensive processing and limitations in recycling, industrial composting and biodegradability. Despite advancements, the practice of greenwashing is prevalent. This chapter discusses ongoing challenges in achieving sustainable additive manufacturing and calls for a balanced approach integrating technological processes with environmental responsibility to shape a sustainable future for 3D printing.

Keywords Sustainability · 3D printing · Additive manufacturing · Life cycle assessment

1 3D Printing Technologies and Common Materials

3D printing technology originated with Charles Hull's invention of stereolithography in the 1980s [1]. This marked the beginning of Additive Manufacturing (AM) and was followed by innovative methods like Fused Deposition Modeling (FDM) and Selective Laser Sintering (SLS). These techniques enabled layer-by-layer object creation. Acrylonitrile Butadiene Styrene (ABS) plastic was a popular choice of

R. Sodhi (✉)
Orproject, New Delhi, India
e-mail: rajat@orproject.com

S. Qin
University of Cincinnati, Cincinnati, OH, USA

© The Author(s), under exclusive license to Springer Nature Singapore Pte Ltd. 2026

C. Klemmt and G. Beaucage (eds.), *Biodegradable 3D-Printing*, The Materials Research Society Series, https://doi.org/10.1007/978-981-95-4025-9_2

material for FDM, while Nylon was used for SLS [2]. While ABS and Nylon are known for their strength, they are petrochemical-based products derived from fossil fuels and are non-biodegradable. This contributes to the accumulation of plastic waste in the environment, a problem known for many years [3].

To address this, Polylactic Acid (PLA), a bioplastic made from corn or sugarcane, has become a popular alternative to ABS due to its renewability and compostability. However, PLA requires specific industrial composting conditions, which are often not accessible, and its source materials compete with food crops, encouraging deforestation and intensive agricultural practices. This makes PLA less than ideal for sustainability [4].

Metals like stainless steel, aluminum and titanium have revolutionized manufacturing in aerospace, automotive and medical fields due to their precision, corrosion-resistant properties and strength. Innovations like Selective Laser Melting (SLM), direct metal laser sintering and Electron Beam Melting (EBM) enable complex, high strength and bespoke components. However, 3D printing metal is highly energy-intensive and relies on non-renewable resources [5].

Recently, cementitious materials like concrete, Geopolymer Concrete and High-Performance Concrete (HPC) have expanded 3D printing applications in the construction industry, reducing waste and enabling complex forms while improving construction timelines. These materials use processes like layered extrusion, Binder Jetting and Slip Forming. However, these materials involve high embodied energy that undermines their sustainability [6].

Current exploration of natural materials such as clay, soil and plant-based materials shows an interest in advancing 3D printing techniques to develop novel solutions that are sustainable and biodegradable. These pioneer technological innovation with traditional materials, opening a new way for sustainable manufacturing of customized designs [7].

2 Evaluation Methods

This study compares information from existing literature focusing on the Life Cycle Assessment (LCA) of materials and their 3D printing processes with data compiled from several sources.

2.1 Life Cycle Assessment

LCA is an all-encompassing technique that meticulously evaluates the ecological ramifications of a product system over the entirety of its life cycle. Beginning with extraction of these raw materials and subsequent production processes, it examines a product system's life through its utilization phase and eventual disposal. The assessment is a compilation of the input, output and environmental impacts of a product through the entirety of its life cycle [8]. These systematic quantifications can guide innovations in 3D printing and manufacturing processes to mitigate their environmental and energy impacts.

2.1.1 Raw Material Extractions

This section examines the procurement of raw materials for 3D printing, assessing impacts like soil degradation, air and water pollution, habitat destruction, biodiversity loss and high energy use [8].

2.1.2 Manufacturing and Processing

Raw materials undergo various processing stages to become usable 3D printing filaments, powders and consumables. Energy intensity is measured for processing, material refinement, synthesis and printing. For example, ABS requires refining and polymerization, while PLA involves fermentation and chemical reactions [9]. Energy use varies by printing technology and material [10].

2.1.3 Transportation

The transportation of materials from production to end-user involves logistics networks, fuel consumption and special storage requirements. For example, filaments used in FDM are susceptible to moisture and temperature changes [10]. As the focus of this study is on manufacturing methods rather than products, we will not look in detail at the effects of transportation.

2.1.4 Usage

This section evaluates the performance of the 3D printed objects in applications like prototyping and construction while profiling energy consumption and waste generation [10]. Due to the variety of products produced by 3D printing and their varying uses, we will not look in detail at the effects of usage.

2.1.5 Waste Disposal

LCA examines waste generation and management of discarded materials, exploring the potential for reducing waste, reusing materials, recycling, industrial composting, incineration for energy and landfilling. For example, PLA composts only under very optimal conditions, while ABS breaks down into microplastics [11].

Biodegradation: Biodegradation involves natural processes by microorganisms, such as bacteria and fungi, into benign components under suitable conditions over time. Beyond the question of whether a material is biodegradable is also the question of the degradation's duration. Even natural materials such as wood may take many years to biodegrade [12].

Industrial Composting: Composting is a conscientiously managed and expedited process that accelerates biodegradation under controlled conditions of humidity, temperature, carbon-to-nitrogen ratio and oxygen [4]. For example, PLA breaks down into carbon dioxide, water and non-toxic byproducts in industrial facilities, but their improper disposal leads to microplastics leaching into the environment [13].

2.2 *Greenwashing*

Greenwashing involves misleading claims about environmental friendliness, leading to increased resource consumption and undermining genuine sustainability efforts [14]. In the 3D printing industry, companies often label their products as "eco-friendly" without evidence of reduction in waste, energy consumption, emissions or biodegradability. For example, PLA is often marketed as biodegradable, but its breakdown requires specialized industrial composting facilities which are not widely accessible [4].

While 3D printing is promoted for its potential sustainability benefits, like minimizing material waste and recycled input materials, these benefits are not always fully realized. For instance, the metal 3D printing industry generates non-recyclable scrap powder, and its energy-intensive processes are overlooked in sustainability claims [5].

These insights call for a comprehensive LCA of 3D printing's environmental impacts and a cautious approach towards the green claims made by industry.

2.3 *Bioplastics*

Bioplastics are a diverse group of polymers that are derived from renewable resources and categorized as bio-based and biodegradable [11]. Bio-based bioplastics, like those from corn or starch may not biodegrade, while biodegradable bioplastics may be made using fossil fuels. For instance, PLA is produced via fermentation of plant sugars into lactic acid and synthesized via monomer polymerization [11]. It is both bio-based and can be broken down using industrial composting and is commonly used as 3D printing filaments and compostable packaging. PLA is also common in medical sutures that dissolve safely in the body [15]. Bioplastics have received considerable attention in recent years due to their potential to alleviate environmental problems associated with conventional plastics which are derived from non-renewable fossil fuels and do not biodegrade [14].

2.4 *Waste in Production*

Private 3D printing is a creative outlet for hobbyists and uses PLA or ABS for printing. The scale of these prints is small and individuals purchase relatively small quantities. However, private 3D printing can be surprisingly wasteful. Failed prints, support structures and leftover material stuck to the print contribute to production waste. Most individuals do not have access to industrial composting facilities or systems to recycle or repurpose waste, so it ends up in the trash, contributing to environmental pollution [10].

Commercial 3D printing, used in aerospace, medical and automotive sectors, is more efficient at employing optimization software to minimize material use and recycling systems for scrap [10].

3 Evaluation

This chapter evaluates the life cycle assessment of plastics (PLA/ABS), metals (aluminum/stainless steel), cementitious (concrete/geopolymer concrete/HPC) and natural materials (clay/soil) used in 3D printing. The evaluation uses LCA conducted by different researchers, assessing each material with a focus on raw material extraction, manufacturing and processing, and waste disposal.

3.1 *Polymers*

Polymers are versatile materials for 3D printing due to their ease of use and affordability for rapid prototyping. Stereolithography (SLA) uses photopolymer resins that harden upon exposure to specific light wavelengths [10]. ABS and PLA are prominent for their strength, flexibility and printability. While ABS is used industrially for durability, PLA is popular for consumer-grade printers due to its lower printing temperature and perceived sustainability [16].

3.1.1 ABS

ABS is a petrochemical-based thermoplastic and is used for its strength and heat resistance, found in applications such as automotive components or LEGO bricks [17].

Table 1 Life cycle assessment data for ABS based on [9]

Impact categories	Unit	Raw material extraction	Manufacturing	Recycling
Climate change	Kg CO_2-Eq	2.39E−01	8.33E−01	4.55E+00
Fossil depletion	Kg oil-Eq	1.12E−01	1.92E−01	1.05E+00
Fresh water ecotoxicity	Kg 1,4−DCB-Eq	4.31E−04	2.28E−04	1.24E−03
Human toxicity	Kg 1,4-DCB-Eq	3.48E−03	2.17E−02	1.19E−01
Ozone depletion	Kg CFC-11-Eq	2.65E−09	1.38E−08	7.56E−08
Particulate matter formulation	Kg PM10-Eq	2.38E−04	2.49E−03	1.36E−02
Terrestrial acidification	Kg SO_2-Eq	6.69E−04	5.18E−03	2.83E−02
Water depletion	m^3	2.15E−04	2.53E−03	1.38E−02

Raw Material Extraction

ABS production starts with the extraction of crude oil and natural gas, which is refined into propylene, ethylene and butadiene. This energy intensive process emits 0.239 kg CO_2-eq/kg and contributes to habitat destruction [9]. Emulsion polymerization for ABS synthesis emits 0.833 kg CO_2-eq/kg (Table 1).

Manufacturing and Processing

ABS filaments are used in Fused Deposition Modeling. Energy consumption varies by printer efficiency and is within 0.1–0.5 kWh/kg of printed ABS [9]. Waste from failed prints and support is common [10].

Waste Disposal

ABS is non-biodegradable, persisting in landfills and fragmenting into microplastics, contributing to marine pollution [3]. Recycling is energy-intensive—4.55 kg CO_2-eq/kg, therefore limiting its sustainability [9].

3.1.2 Polylactic Acid

PLA is an industrially compostable thermoplastic derived from renewable resources. It has low printing temperatures and minimal warping, making it ideal for novice 3D printing enthusiasts as an alternate to ABS [16].

Table 2 Life cycle assessment data for PLA based on [9]

Impact categories	Unit	Raw material extraction	Manufacturing	Recycling
Climate change	Kg CO_2-Eq	1.95E−01	6.61E−01	2.87E+00
Fossil depletion	Kg oil-Eq	5.27E−02	1.52E−01	6.60E−01
Fresh water ecotoxicity	Kg 1,4-DCB-Eq	1.61E−03	1.81E−04	7.85E−04
Human toxicity	Kg 1,4-DCB-Eq	1.49E−02	1.73E−02	7.49E−02
Ozone depletion	Kg CFC-11-Eq	1.39E−08	1.10E−08	4.77E−08
Particulate matter formulation	Kg PM10-Eq	4.05E−04	1.98E−03	8.59E−03
Terrestrial acidification	Kg SO_2-Eq	1.28E−03	4.11E−03	1.78E−02
Water depletion	m^3	2.41E−02	2.01E−03	8.71E−03

Raw Material Extraction

PLA production involves fermenting plant-based sugars into lactic acid and requires agricultural inputs like land, water and fertilizer. It emits 0.195 kg CO_2-eq/kg and is lower than ABS as it used renewable feedstocks [9]. Fermentation and polymerization consume 50–60% fewer fossil fuels than ABS, emitting 0.661 kg CO_2-eq/kg [9]. Filament production requires additional energy.

Manufacturing and Processing

PLA is processed into filaments for FDM, with printing consuming 0.05–0.3 kWh/kg due to lower melting points [9]. Waste from support and failed prints is common.

Waste Disposal

PLA composts in industrial facilities in 6–24 months but forms microplastics if improperly disposed [13]. Recycling emits 2.87 kg CO_2-eq/kg and industrial composting requires energy-intensive conditions [4]. As industrial composting facilities are sparse, most PLA waste ends up in solid waste landfills (see Table 2).

Table 3 Energy (MJ/kg) in raw material extraction and manufacturing for different AM processes. Averaged data based on [5]

Energy (MJ/kg)		Material extraction	Manufacturing	Total
Nylon	SLS	129.1	97.96	227.06
Stainless steel	SLM	55.1	205.99	261.09
Stainless steel	DED	55.1	13.09	68.19
Aluminum	SLM	206.5	612.75	819.25
Aluminum	DED	206.5	51.8	258.3
Titanium	EBM	572.25	225.7	797.95
Titanium	DED	572.25	209.4	781.68
PLA	3DP	55.8	38.09	93.89
ABS	3DP	93.04	85.19	178.23
Nylon	3DP	129.1	48.19	177.295
Concrete	3DP	1.11	1.02	2.14

3.2 *Metals*

Stainless steel and Aluminum are key metals known for their strength and corrosion resistance. The development of SLM and EBM technologies opens the space for using these materials for 3D printing.

3.2.1 Stainless Steel

Stainless steel is durable and corrosion-resistant and is a popular choice in the automotive industry and to make industrial parts.

Raw Material Extraction

As an alloy of iron, chromium and nickel require the mining of these separate elements. The extraction of iron, chromium and nickel consumes 20–40 MJ/kg and causes habitat destruction, emitting 1.5 kg CO^2-eq/kg (Table 3) [5].

Manufacturing and Processing

SLM and direct metal laser sintering (DMLS) consume 80–150 MJ/kg, with post-processing (stress-relieving, polishing) adding 10–20 MJ/kg [5]. Powder waste is significant. Stainless steel printing consumes 40–80 kWh/kg (Table 3) [5].

Waste Disposal

Recycling requires 10–20 MJ/kg, with minimal losses [5]. It is non-biodegradable.

3.2.2 Aluminum

Aluminum's lightweight and high strength make it an ideal material to 3D print parts used in aerospace and automotive industries. Like stainless steel, aluminum is processed using SLM or DMLS methods in 3D printing.

Raw Material Extraction

Aluminum extraction from bauxite via the Bayer and Hall-Héroult process consumes 15 kWh/kg and emits 1.23 kg carbon dioxide equivalent per kilogram produced (Table 3) [5].

Manufacturing and Processing

Printing consumes 30–60 kWh/kg [5]. Supports generate recyclable waste [5].

Waste Disposal

Whilst aluminum is non-biodegradable, it is easily recyclable and avoids long-term pollution. Recycling uses 5% of primary production energy 0.75 kWh/kg, making it cost-effective [5].

3.3 Cementitious Materials

The advent of 3D printing technology has revolutionized various sectors, including construction, by introducing innovative materials like concrete, geopolymer concrete and high-performance concrete (HPC). The technique works by layer deposition of concrete to create three dimensional structures.

Geopolymer concrete emerges as a sustainable alternative to traditional concrete, utilizing industrial by-products like fly ash and slag, which significantly reduce the carbon footprint [18].

HPC is known for its superior strength and durability and incorporates materials like silica fume and superplasticizers to achieve enhanced properties. In 3D printing, HPC can create structures that withstand extreme conditions and are crucial for infrastructural resilience [19].

Table 4 Life Cycle Assessment data for cementitious and earth materials based on [23]

Material	Source	Embodied carbon (kg CO_2-eq/kg)
3D printed concrete	[20, 24]	0.208
3D printed earth (WASP)	[25]	0.077
3D printed earth (PM)	[23]	0.006

3.3.1 Concrete

Traditional concrete is composed of cement, aggregates and water. It is widely used for 3D-printed buildings.

Raw Material Extraction

Limestone quarrying for cement emits 0.8 kg CO^2-eq/kg with aggregates adding minimum impact. Quarrying disrupts ecosystems. Cement production is energy-intensive at 4–5 MJ/kg [20].

Manufacturing and Processing

Concrete printing for housing or models consumes 0.5–2 kWh/m^3 (Table 4) [20].

Waste Disposal

To some extent, concrete waste from demolition can be recycled as an aggregate and requires 0.5–1 MJ/kg. However, most of the construction demolition waste contributes to landfills.

3.3.2 Geopolymer Concrete

Geopolymer concrete uses industrial by-products like fly ash and slag thereby reducing emissions [18, 21, 22].

Raw Material Extraction

Fly ash and slag collection requires 0.1–0.5 MJ/kg with alkaline activators adding chemical impact. Geopolymer production emits 0.2–0.4 kg CO_2-eq/kg.

Manufacturing and Processing

3D printing using geopolymer concrete adds 1–2 MJ/kg. Curing conditions increase energy use. Geopolymer printing consumes 0.5–2 kWh/m^3 [18].

Waste Disposal

Recycling is energy intensive and uses 1–2 MJ/kg.

3.3.3 High Performance Concrete

Known for its superior strength and durability, HPC is used in 3D printing for constructing elements that require high performance, such as in extreme weather conditions or heavy load-bearing structures [19].

Raw Material Extraction

Like concrete, HPC uses high-quality cement and silica fume, emitting 0.9 kg CO_2-eq/kg. Admixtures add chemical impacts. HPC production consumes 5–6 MJ/kg [20].

Manufacturing and Processing

HPC printing adds 1–2 MJ/kg. Precision mixing increases energy. When used for bridges and architectural elements, HPC printing consumes 0.5–2 kWh/m^3 [20].

Waste Disposal

The waste disposal of HPC is like that of traditional concrete, but due to its enhanced properties, it might be more challenging to recycle. The high strength of HPC can make processing and crushing for recycling more energy intensive and uses 1–2 MJ/kg [20].

3.4 Clay and Soil

Clay and Soil are natural materials that offer sustainable options for 3D printing with uses in ceramics, art and architecture. These materials are mostly extruded layer-by-layer using a desktop clay extruder or a robotic arm.

3.4.1 Clay and Soil

Clay is abundant and used in architectural applications.

Raw Material Extraction

Open-pit mining of clay requires 0.2–0.5 MJ/kg and causes land degradation. Soil extraction uses 0.1–0.3 MJ/kg, with risks of erosion and biodiversity loss. Additives such as lime are the main cause of CO_2 emissions [23].

Manufacturing and Processing

Clay is processed into printable pastes, consuming 0.5–1 MJ/kg. Extrusion printing adds minimal energy of 0.1–0.5 kWh/m^3. Used for sustainable housing such as the TECLA project [26], clay printing consumes 0.1–0.5 kWh/m^3 (Table 3) [23].

Waste Disposal

While not biodegradable, the waste is harmless and occurs abundantly in nature. Recycling as raw material is feasible [23].

4 Discussion

The evolution of 3D printing materials from petrochemical-based plastics like ABS to bioplastics, metals, cementitious and natural materials reflect technological advancements and a commitment to sustainability. The shift to PLA, a renewable and industrially compostable bioplastic, significantly reduces the carbon footprint compared to ABS, especially when renewable energy is used for production [9]. However, PLA's reliance on industrial composting facilities, which are not widely accessible, leads to solid waste or microplastic pollution if mismanaged, highlighting the need for better waste management infrastructure [13]. While PLA represents progress, natural and plant-based materials like cellulose-based bio composites offer greater potential for circularity due to their biodegradability in natural environments and lower land-use impact.

Metal 3D printing, using stainless steel and aluminum, enables complex geometries and reduces material waste compared to traditional subtractive methods [5]. However, the energy-intensive production of metal powders and printing processes remains a challenge. Developing recyclable metal powders and energy-efficient technologies could enhance sustainability.

Cementitious materials like concrete, geopolymer concrete and HPC have significant environmental impacts due to raw material extraction. Geopolymer concrete uses industrial by-products and offers a lower carbon footprint. However, waste disposal and recycling remain an issue.

Clay, soil, and plant-based materials [27] are promising for sustainable 3D printing. Their low embodied energy, local availability and biodegradability align with circular economy principles. However, extraction must be managed to avoid ecosystem disruption and standardized construction frameworks are needed to ensure scalability and safety. Furthermore, materials that can biodegrade, do biodegrade. Product durability is therefore a concern for many biodegradable or soil materials, while also issues persist in the processability of novel and experimental materials and production techniques.

5 Conclusion

We have compiled information on the sustainability of various 3D printing methods and materials. Major concerns exist on the one hand in the energy use for the raw-material extraction, processing, and manufacturing, especially using cementitious materials and metals, and on the other hand in the waste management and limited or non-existing circularity, especially in the case of concrete and polymers. Plant-based materials, such as cellulose or hemp-based bio composites, may hold immense potential due to their renewability, biodegradability and minimal environmental impact. Prior to the polymer explosion, most consumer products were produced from plant fiber, while early polymers such as cellophane were made from bioplastics. These materials could redefine additive manufacturing by enabling fully circular systems, particularly in construction and consumer goods, where local sourcing reduces transportation emissions. However, further research is required in the development of feasible and scalable manufacturing.

The future of 3D printing lies in balancing environmental sustainability, technological advancement and ethical practices. Continued research into plant-based materials alongside energy-efficient processes can minimize ecological footprints of 3D printing while maximizing sustainable possibilities and creating economies of circularity.

References

1. Hull, C.: The birth of 3D printing. Res. Technol. Manag. **58**, 25–30 (2015)
2. Parupelli, S., Desai, S.: A comprehensive review of additive manufacturing (3D printing): processes, applications and future potential. Am. J. Appl. Sci. **16** (2019)
3. Geyer, R., Jambeck, J., Law, K.: Production, use, and fate of all plastics ever made. Sci. Adv. **3**, e1700782 (2017)

4. Mistry, A., Kachenchart, B., Pinyakong, O., Assavalapsakul, W., Jitpraphai, S., Somwangthanaroj, A., Luepromchai, E.: Bioaugmentation with a defined bacterial consortium: a key to degrade high molecular weight polylactic acid during traditional composting. Bioresour. Technol. **367**, 128237 (2023)
5. Kokare, S., Oliveira, J., Godina, R.: Life cycle assessment of additive manufacturing processes: a review. J. Manuf. Syst. **68**, 536–559 (2023)
6. Nematollahi, B., Xia, M., Sanjayan, J.: Current progress of 3D concrete printing technologies. In: ISARC Proceedings of the International Symposium on Automation and Robotics in Construction, vol. 34 (2017)
7. Carcassi, O., Ben-Alon, L.: Additive manufacturing of natural materials. Autom. Constr. **167**, 105703 (2024)
8. Hauschild, M., Rosenbaum, R., Olsen, S.: Life Cycle Assessment 2018. Springer, Cham (2018)
9. Kumar, R., Sharma, H., Saran, C., Tripathy, T., Sangwan, K., Herrmann, C.: A comparative study on the life cycle assessment of a 3D printed product with PLA, ABS & PETG materials. Procedia CIRP. **107**, 15–20 (2022)
10. Gibson, I., Rosen, D., Stucker, B., Khorasani, M.: Additive Manufacturing Technologies, vol. 17. Springer, Cham (2021)
11. Sudesh, K., Iwata, T.: Sustainability of biobased and biodegradable plastics. Clean–Soil Air Water. **36**, 433–442 (2008)
12. Kim, Y., Singh, A.: Micromorphological characteristics of wood biodegradation in wet environments: a review. IAWA J. **21**, 135–155 (2000)
13. Liu, R., Liang, J., Yang, Y., Jiang, H., Tian, X.: Effect of polylactic acid microplastics on soil properties, soil microbials and plant growth. Chemosphere. **329**, 138504 (2023)
14. Thompson, R., Moore, C., Vom Saal, F., Swan, S.: Plastics, the environment and human health: current consensus and future trends. Philos. Trans. R. Soc. B. **364**, 2153–2166 (2009)
15. Barcena, A., Ravi, P., Kundu, S., Tappa, K.: Emerging biomedical and clinical applications of 3D-printed poly(lactic acid)-based devices and delivery systems. Bioengineering. **11**, 705 (2024)
16. Tümer, E., Erbil, H.: Extrusion-based 3D printing applications of PLA composites: a review. Coatings. **11**, 390 (2021)
17. Andrady, A., Neal, M.: Applications and societal benefits of plastics. Philos. Trans. R. Soc. B. **364**, 1977–1984 (2009)
18. Yao, Y., Hu, M., Di Maio, F., Cucurachi, S.: Life cycle assessment of 3D printing geopolymer concrete: an ex-ante study. J. Ind. Ecol. **24**, 116–127 (2020)
19. Gosselin, C., Duballet, R., Roux, P., Gaudillière, N., Dirrenberger, J., Morel, P.: Large-scale 3D printing of ultra-high performance concrete–a new processing route for architects and builders. Mater. Des. **100**, 102–109 (2016)
20. Roux, C., Kuzmenko, K., Roussel, N., Mesnil, R., Feraille, A.: Life cycle assessment of a concrete 3D printing process. Int. J. Life Cycle Assess. **28**, 1–15 (2023)
21. Jin, W., Roux, C., Ouellet-Plamondon, C., Caron, J.-F.: Life cycle assessment of limestone calcined clay concrete: potential for low-carbon 3D printing. Sustain. Mater. Technol. **41**, e01119 (2024)
22. Khan, S., Jassim, M., Ilcan, H., Sahin, O., Bayer, İ., Sahmaran, M., Koc, M.: 3D printing of circular materials: comparative environmental analysis of materials and construction techniques. Case Stud. Constr. Mater. **18**, e02059 (2023)
23. Curth, A., Pearl, N., Castro-Salazar, A., Mueller, C., Sass, L.: 3D printing earth: local, circular material processing, fabrication methods, and life cycle assessment. Constr. Build. Mater. **421**, 135714 (2024)
24. Papachristoforou, M., Mitsopoulos, V., Stefanidou, M.: Evaluation of workability parameters in 3D printing concrete. Procedia Struct. Integr. **10**, 155–162 (2018)
25. Ferretti, E., Moretti, M., Chiusoli, A., Naldoni, L., De Fabritiis, F., Visonà, M.: Mechanical properties of a 3D-printed wall segment made with an earthen mixture. Materials. **15**, 438 (2022)

26. Di Salvo, S.: Insights into new trends and contemporary challenges in 3D printing in architectural sector. Eng. Innov. **8**, 43–61 (2024)
27. Klemmt, C., Aghaei Meibodi, M., Beaucage, G., McGee, W.: Large-scale robotic 3D printing of plant fibre and bioplastic composites. In: Proceedings of the eCAADe (2022)

Rajat Sodhi is an architect and founding partner at Orproject, a research based design practice with offices in New Delhi, London, and Cincinnati. Holding a diploma and DRL master's from the Architectural Association in London, Sodhi has worked with renowned firms like Herzog and de Meuron and Foster + Partners. His work focuses on biomimetic and computational design, integrating nature-inspired geometries with ecological solutions. Notable projects include the Bubbles, addressing urban air quality, and Vana, a paper forest installation. Sodhi's innovative approach has earned recognition, including Orproject's inclusion in Architectural Digest's AD100 list.

Shuo Qin is a six-year doctoral student in the architecture program at the University of Cincinnati College of Design, Architecture, Art, and Planning. His current field research focuses on how brutalism and Urban Renewal reshaped North American Chinatown, specifically Sacramento Chinatown, San Francisco Chinatown, and Washington DC Chinatown. From those Chinatowns, he put the lens on the history of immigrants, and the occupation of architecture(brutalism) before and after the design. He also demonstrates the aesthetics of Chinese elements within the architecture. His secondary research focuses on the study of the comparison of the informal settlement between Latin America and China.

Chapter 3
Review of Biodegradable Materials in Additive Manufacturing

Gregory Beaucage and Christoph Klemmt

Abstract Additive manufacturing involves production of customized parts of small quantity that inherently involves creation of substantial waste through failed trials. Further, some fabrication processes involve substantial amounts of unused material. For most of the processes, materials are substantially degraded in the process and blended with multiple components inhibiting reuse and recycling. For these reasons, the use of biodegradable materials in additive manufacturing may address this solid waste problem. Many materials that claim biodegradability such as polylactic acid are not recyclable in the current US society. Some materials like polyhydroxybutyrate are cost prohibitive. Biomaterials such as cellulose/wood and chitosan/chitin offer substantial potential at low cost and with real circularity. Biodegradable, adaptable, animate materials containing living organisms offer a window into the future of biosourced, biodegradable, and bioactive advanced manufacturing.

Keywords Biodegradable materials · Advanced manufacturing · Additive manufacturing · 3D printing

1 Introduction

Additive manufacturing (AM) couples conventional materials processing methods such as extrusion, photopolymerization, powder processing, reactive extrusion and casting with computer-controlled positioning of a part or of the processing device with the intent of fine control over the three-dimensional assembly towards customizable or unique forms usually in a layer-by-layer process. Additive manufacturing allows for the rapid development of prototypes and custom parts, on-site, and with

G. Beaucage
Chemical and Materials Engineering, University of Cincinnati, Cincinnati, OH, USA

C. Klemmt (✉)
College of Design, Architecture, Art and Planning, University of Cincinnati, Cincinnati, OH, USA
e-mail: christoph@orproject.com

© The Author(s), under exclusive license to Springer Nature Singapore Pte Ltd. 2026
C. Klemmt and G. Beaucage (eds.), *Biodegradable 3D-Printing*, The Materials Research Society Series, https://doi.org/10.1007/978-981-95-4025-9_3

on-demand assembly. A byproduct of the simplicity and low cost of AM is the generation of substantial amounts of waste material which exacerbates an existing crisis in solid waste management especially for large-scale applications and/or for single-use items. Reuse of waste AM material is usually not possible since extruded materials or cured/reacted materials generally lose their processability, are of a lower quality, and may contain complex mixtures that cannot be reformulated. Recycling or downcycling of AM materials largely does not exist, except for potential use as fuel or in chemical depolymerization processes which are largely not realistically available especially for non-uniform AM waste streams. Upcycling of AM materials has not been significantly explored but does not seem to offer promise. An option that has been implemented with polylactic acid and blends is use of "compostable" materials in AM. This holds the potential of significantly reducing the solid waste burden in AM processes. However, consideration must be made as to the realistic possibility of biodegradation in the existing waste streams; degradation byproducts which may be toxic or lead to eutrophication; environmental hazards involved in the production of biodegradable materials such as the wood farming/paper pulping process; biological hazards such as mold and fungus; realistic processing requirements and engineering design principles for materials such as wood chip slurries; mechanical and physical properties of biodegradable substitutes; lifetime performance for these materials; as well as a general lifecycle analysis of their environmental impact to highlight a few areas. We can consider several select examples where active research is being done to expand the use of AM towards materials that have a lower environmental burden. Table 1 lists the biodegradable materials considered in this chapter.

In addition to the solid waste issue, biosource, biodegradable materials for AM can be designed with living cells such as algae, bacteria and fungi, as well as human cells. These materials promise a new field, *animate materials*, that can dynamically adapt to their environment based on environmental conditions.

2 Defining Biodegradable Materials

It should be clear that "biodegradable" has been abused as a descriptive term with the implication that biodegradable materials will solve and not create environmental problems and that the possibility of biodegradation implies functional biodegradation in normal use. Non-biodegradable materials are, in their essence, generally inert, non-reactive, stable materials which were seen as positive features during the petrochemical explosion of the 1950s–1970s driven by technological developments necessitated by WWII in the 1940s. In this boom period little thought was given to life-cycle assessment for materials whose lifetime was essentially infinite on a human timescale. A few people could see an emerging problem with plastics waste and industrial scale pollution, but they were largely ignored. An engineered lifetime for materials is a topic that is still poorly understood and is a requirement for truly biodegradable materials. This requires a life-time assessment with the intention of understanding unforeseen consequences of mass-production and use by normal

Table 1 Biodegradable materials considered in this chapter

Material	Acronym	Source	Biodegradability	Environmental impact	Approximate cost $/kg (compare to 1.5 for PE/PP/PET)
Polylactic acid	PLA	Fermentation of food crops after industrial purification, polymerization of lactide in industrial process	Industrially compostable, effectively not biodegraded in the US	High potential for microplastics production, non-toxic degradation products	3
Polyhydroxybutyrate	PHB	Fermentation of food or waste streams, PHB is a direct fermentation product	Compostable and highly biodegradable	No microplastics, non-toxic degradation products	15
Polyhydroxybutyrate-co-valerate	PHBV	Like PHB	Like PHB	Like PHB	15
PLA blends and composites	PLA	Like PLA	Like PLA	High potential for microplastics, non-toxic degradation products	3
PHB blends and composites	PHB	Like PHB	Like PHB	Potential for microplastics production depending on additive	15
Polycaprolactone	PCL	Petrochemical	Compostable non-toxic byproducts	Non-toxic degradation	10
Polybutylene-adipate-co-terephthalate	PBAT	Petrochemical	Industrially compostable, effectively not biodegraded in the US	Degradation produces terephthalic acid, an aromatic hydrocarbon	4

(continued)

Table 1 (continued)

Material	Acronym	Source	Biodegradability	Environmental impact	Approximate cost $/kg (compare to 1.5 for PE/PP/PET)
Polyvinyl alcohol	PVA or PVOH	Petrochemical	Compostable, degradation leads to glycolic acid, acetic acid,	Non-toxic degradation byproducts	4
Polyethylene oxide	PEO or PEG	Petrochemical	Non compostable	Remains in the environment as a water-soluble polymer	5
Pluronic	PEO-co-PPO	Petrochemical	Like PEO	Like PEO	100
Wood chips	–	Biosource	Compostable	Can be obtained from byproducts, limited environmental impacts	0.05
Wood pulp	–	Biosource with extensive chemical processing	Compostable	Chemical processing can impact water resources	1
Cellulose	–	Like pulp	Like pulp	Like pulp	2
Nano-cellulose	NC, CNF, CNC, BNC	Like pulp	Like pulp	Like pulp	200
Lignin	–	Chemical Process from wood chips	Essentially non-compostable with aromatic byproducts	Aromatic byproducts, largely non-compostable	0.5
Starch	–	Separation from food crops	Compostable with	No dangerous byproducts	0.5

(continued)

Table 1 (continued)

Material	Acronym	Source	Biodegradability	Environmental impact	Approximate cost $/kg (compare to 1.5 for PE/PP/PET)
Chitin	–	Produced by an industrial chemical separation process from shells, water soluble	Compostable	Produces saccharides and glucosamine both non-toxic	10
Chitosan	–	Like chitin but with deacetylation which makes product hydrophobic	Compostable	Produces saccharides and glucosamine both non-toxic	100
Fungus-like additive material	FLAM	Combination of cellulose fiber and chitin	Compostable	Non-toxic byproducts same as chitin and cellulose	7
Sodium alginate	–	Made from brown seaweed by a separation process	Compostable	Non-toxic byproducts	10
Fucoidan	–	Made from brown seaweed	Compostable	Non-toxic byproducts	10
Carrageenan	–	Made from red seaweed by a separation process	Compostable	Non-toxic byproducts	10
Agarose	–	Made from red seaweed by a separation process	Compostable	Non-toxic byproducts	100

(continued)

Table 1 (continued)

Material	Acronym	Source	Biodegradability	Environmental impact	Approximate cost $/kg (compare to 1.5 for PE/PP/PET)
Hyaluronic acid	–	Made by bacterial fermentation and separation	Compostable	Non-toxic byproducts	10
Collagen	–	Extracted from animal tissue waste	Compostable	Non-toxic byproducts	10
Gelatin	–	Extracted from animal tissue waste	Compostable	Non-toxic byproducts	5

people in normal lived environments as well as an assessment of the expected time of use and time for degradation [1].

Biodegradation of composite materials and materials that show partial degradation can lead to the production of micro and nano plastic particles. This is important to materials like PLA blends which are common AM filaments. Further, the degradation byproducts of some biodegradable materials such as PBAT can be environmentally harmful, as PBAT releases terephthalic acid, an aromatic hydrocarbon during biodegradation.

3 Thermoplastics for AM

Currently two main categories of thermoplastic materials are marketed for compostable AM, polylactic acid (PLA) and polyhydroxyalkanoates (PHAs), particularly polyhydroxybutyrate (PHB). PLA is synthesized from a cyclic monomer, lactide, which is produced from lactic acid coming from fermentation of sugar from corn, sugar cane and other food sources in an industrial chemical processing operation. PLA can degrade biologically at high temperatures under controlled conditions in certain types of industrial composters. These facilities are for the most part not available in the US, of 87 industrial composting facilities only about 4 accept some types of sorted, unblended PLA plastics. Sorting of PLA from other plastics is a major issue that relies on labeling which is absent from AM waste streams. PLA is a brittle plastic, so it is usually blended with synthetic plastics, polybutylene adipate terephthalate (PBAT), starch or plant fibers, polyethylene terephthalate (PET) or polypropylene (PP). Since PLA is not degradable under normal composting conditions, the addition of degradable materials, PBAT and starch/fibers, results in the production of crystalline microplastic PLA particles which can exacerbate the waste problem. PET and PP are not degradable, addition to PLA makes the compound non-degradable even in one of the four US-sites suitable for industrial PLA biodegradation. The materials are not sufficiently labeled to distinguish between non-degradable and degradable PLA blends. In terms of a lifecycle assessment, the energy use when comparing PLA production to PET production is about equivalent as are the greenhouse gas emissions, ignoring the energy needed to raise the food crops that are a feed stock to fermentation. PLA production introduces stress on the food economy and results in significant pesticide, fertilizer, water, and land use contamination and would appear to compete with food that might be used to feed people.

PHAs and PHB are produced by organisms, generally by bacterial fermentation, using a variety of feedstocks including food and waste streams such as sugar, starch, and agricultural residue. Through transgenic engineering, soybean, cactus, switchgrass, tobacco, model plants and algae have been modified to directly produced PHAs removing the fermentation stage of production. Large scale production from soybeans was seen as a commercially viable approach by Procter and Gamble in the 1990s but was not fully implemented. Separation of PHAs from the plant matter is an added processing step. Displacement of food crops and food land use are issues with agricultural production of PHAs, though less when compared with PLA since many alternative sources exist, for instance use of suboptimal agricultural land in arid regions. Energy consumption for PHAs is about half of petrochemical plastics like PET with about 1/6 of the greenhouse gas emissions when produced from sugarcane. PHB is biodegradable under any conditions and does not produce microplastics. PHB is a brittle semi-crystalline polymer. To toughen PHB it is usually produced as a co-polymer with polyhydroxy valerate (PHBV) and other aldonates which reduces the crystallinity making the plastic easier to process and more flexible. The impact of bacterial copolymerization on biodegradation can vary. Reduction of crystallinity leads to superior processing and mechanical properties and biodegradation rate in normal compost is increased since crystals block access to the polymer chains, but due to increased hydrophobicity of longer alkane side groups, degradation in an aqueous environment such as the ocean is slowed somewhat. Microplastics are not produced by any PHAs since they completely and rapidly decay. PHB and copolymers seem to be a useful biodegradable and bio-sourced plastic for AM. Due to the cost of PHAs compared to PLA they are rarely used for AM. About 45% of AM by plastic deposition uses PLA. PHAs are used insignificantly, mainly for medical devices that require biodegradation. A rough cost comparison is PP/PET ~ $1.5/kg, PLA ~ $3/kg, PHAs ~ $15/kg. PLA, PP, and PET have similar processability while PHB has some problems with processing due to low and unstable viscosity and a lower thermal degradation temperature.

Thermoplastic starch (TPS) is starch milled with a plasticizer at moderate temperatures. The plasticizer can be water, glycol or other polyols all of which are biodegradable. Starch, which is crystalline and powdery, is transformed into a viscoelastic amorphous or semi-crystalline material like a polymer melt. Properties can vary significantly depending on the processing. TPS is hydrophilic and loses properties when exposed to humidity, but it can be blended with PLA or PHAs as a composite material with reduced sensitivity to humidity. TPS is similar in price to petrochemical polyolefins like PP. Figure 1 shows a typical commodity use for PHBV.

4 Other Synthetic Biodegradable Plastics for AM

Several fossil-based compostable polymers exist that are used either as an AM resin or in blends with PLA and PHB for melt extrusion AM. Polycaprolactone (PCL) is a biodegradable polyester usually derived from benzene (a constituent of crude oil); polybutylene-adipate-co-terephthalate (PBAT) is a biodegradable random copolymer derived from petroleum that is often used in compostable cling wrap and plastic bags, polyvinyl alcohol (PVA or PVOH) is a petrochemically derived water-soluble synthetic polymer, polyethylene oxide (PEO or PEG) is a polyether derived industrially from petroleum, pluronics are block copolymers of PEO (hydrophilic) and polypropylene oxide (PPO, hydrophobic) which form extrudable hydrogels.

Fig. 1 PHA cutlery from Tiptopak, www.tiptopak.com

5 Wood-based Materials for AM

Wood is composed of cellulose (a linear, rigid-chain, hydrophilic polymer), hemi-cellulose (branched/network polymers with composition like cellulose), lignin (brittle, glassy, crosslinked phenolic) and other minor components. Cellulose is a semi-crystalline polymer which is significantly plasticized by water, hemi-cellulose is amorphous and is also significantly plasticized by water. Wood chips and powder can be directly used in AM, or it can be pulped to enhance processability. Pulping can remove lignin and hemi-cellulose to different extents. Wood itself, is biodegradable and bio sourced, often available as a waste stream from lumber production and paper making. Wood chips, powder, or pulp are processed using water and structural changes during drying are a major consideration. The three types of pulping processes are: mechanical pulping which produces a mixture of cellulose/hemicellulose and lignin used for newsprint and other short-lived products after bleaching; chemical pulping removes most hemicellulose and all lignin (normal non-yellowing paper); semi-chemical pulping removes some lignin (cardboard).

Cellulose is naturally biodegradable, and bio-sourced. However, chemical pulping operations lead to significant energy consumption and environmental impact. Cellulose from chemical pulping is the original source of biobased plastics starting with cellulose nitrate and cellulose acetate. These are non-biodegradable bio sourced thermoplastics.

Current wood composites are 2D materials, including paper, plywood, particle board, fiberboard, and Masonite hardboard. Paper is a 2D material produced from wood after chemical or mechanical pulping that can remove lignin and hemi-cellulose and involves the addition of about 8% starch as a binder. 2D cellulose based structural materials involve layered processing with application of steam and pressure and generally an added binder. The binder is either a synthetic, petrochemical (except in specialty applications) thermoset resin that can carry out condensation reactions with the hydroxyl groups in cellulose and hemi-cellulose, or wood lignin which binds cellulose after steam pressure treatment and expands wood fiber in Masonite. Chemical mechanical pulp includes a heat treatment step with Na_2SO_3 and Na_2CO_3 solution, which breaks down some hemi-cellulose and removes resin. Masonite pulp is a related material produced from pressurized steam followed by rapid depressurization, which explodes wood chips, leading to wood fibers that contain lignin and hemi-cellulose. In Stein-Hall adhesives, which are used in the production of corrugated paper, starch is modified with ionic chemicals that can lead to condensation of the hydroxyl groups. Specifically, borax, caustic (NaOH), sodium carbonate, or sodium silicate modify starch, leading to tunable viscosity and the ability to adhere to cellulose. The condensation process is driven by the removal of water at 122 to 158 °C [2].

There is limited literature describing high wood-content composites with greater than 85% wood content for AM 3D-printing (3DP). Compostable, but not biodegradable, wood-like filaments are commercially available for desktop 3D-printers. These filaments are composed of about 70% polylactic acid (PLA) and 30% wood fiber [3, 4]. The material is industrially compostable and has a density like wood. Biodegradation of the wood is likely to lead to microplastic PLA. A similar composition has been used at a larger scale [5]. Kariz et al. [6] used polyvinyl acetate (PVAc) and urea–formaldehyde (UF) adhesives as binder with fine wood powder to extrude 3D-printed objects. Due to high viscosity, the wood content was limited to 12–25%. The extrudate exhibited about 20% shrinkage upon curing. Early-stage research has investigated small-scale 3DP by extrusion using liquid deposition of fine wood powder with cellulose or lignin binder with 90% wood and up to 100% content of wood products [7, 8]. Shrinkage after curing and drying is on the order of 20%. In contrast, in the work of Klemmt et al. [9] control of material compositions allows for greater control over the shrinkage or expansion of the extrudate.

Henke and Treml [10] further expanded on powder-based 3D printing through investigation of various wood products (sawdust/wood chips) and several types of binding agents (gypsum/cellulose/sodium silicate/cement). This involved a hybrid of the extrusion and powder methods. In this process, the best results occurred with cement as the binder, premixing the chips and cement and deposition followed by application of a water aerosol, resulting in 70% wood content. Zeidler et al. [11] used powder bed 3DP and various binders, including sodium silicate, to bind various bio-based fibers, with up to 94% wood. The best results were found for polyvinyl alcohol. Rosenthal et al. [7] used sawdust and methyl cellulose binder with a dry content of 89% wood, and 100% wood products. The extrusion slurry was composed of 90% wood dust and 10% of a 5% methylcellulose water mixture. Drying produced a 20% reduction in linear dimensions of the printed part.

Various complicated processes have been developed using 3DP of wood derivatives. Hausmann developed a solvent exchange method, where an aqueous cellulose paste is extruded into a complex shape, the solvent is exchanged with an organic solvent, infiltrated with a synthetic monomer, polymerized and dried [12]. This process led to high-resolution, shape-retaining wood-based objects with about 30% wood. Another approach involves direct cryo-writing, where an aqueous suspension of cellulose or other wood-based materials is frozen during 3d printing. The process has been proposed as a method to generate highly oriented fibers [13]. The printed object is placed in a vacuum and freeze dried to remove water by sublimation [8]. Liangbing Hu at the University of Maryland [14] has developed a process using lignin and triblock Pluronic® that can crosslink lignin molecules forming a shear thinning gel in water. The extrudable gel is an effective 3DP slurry and can be used at room temperature. Water is removed from the product by freeze drying, resulting in 90% wood-based 3D objects. None of these more complex approaches for 3DP of wood can be used to produce large-scale objects. Most are too expensive or are not currently possible on a large scale.

The current methods used for large-scale 3DP have several drawbacks for use in moderate to large scale applications like construction including high cost, and use of synthetic hydrocarbon-based materials, producing non-biodegradable parts. Existing methods for 3DP of biodegradable wood materials remains small-scale. Use of commercial filaments, while capable of larger scale, use little wood (30%) and depend on thermoplastic extrusion, which is an energy-consuming high-pressure process that requires sufficient cooling before applying subsequent layers. In contrast, the work of Klemmt et al. [9] provide strong evidence that a new 3DP method has the potential to be both environmentally friendly and economically feasible at large scales.

5.1 Additives for 3DP of Wood

Several bio- and silica-based additives have been proposed to enhance 3DP of wood: starch, agar, and silanes. Like cellulose, starch is a polysaccharide composed of glucose units. In starch, the glucose units are arranged in the same orientation (Amylose $\alpha(1 \rightarrow 4)$ glycosidic bonds), in cellulose the units alternate by 180 degrees ($\beta(1 \rightarrow 4)$ glycosidic bonds) [2, 15]. Starch is commonly used as a binder adhesive for cellulose in the manufacture of paper and cardboard. It is cationic, while cellulose is anionic, so the two can form a coacervate gel. Starch is also used as a glue with gypsum to produce wallboard. Starch can be mixed with borax or sodium carbonate (soda ash) to form an adhesive. Sodium silicate (water glass) can be added as a reinforcement. Several groups have used aqueous starch slurries for 3DP. Theagarajan et al. [16] and Chen et al. [17] recently explored 3DP starch and found viable composition and processing conditions for high-resolution extrusion 3DP. Starch has also been used in binder jet 3DP for pharmaceuticals [18]. Yang et al. [19] used starch as a sacrificial phase in 3DP of ceramic slurries. Agar is another polysaccharide that has been used as 3DP slurry. Holland et al. demonstrated that agar could be 3DP [20]. A group at ETHZ has produced 3DP bacterial colonies using agar [21]. Agar is composed of agarose and agaropectin which are closely related to cellulose and starch in structure. Wang et al. printed tough agar strain sensors by crosslinking 3DP agar gels with calcium salts [22]. Calcium is a bivalent cation that can bind to the anionic hydroxyl groups on agar, cellulose, or starch, leading to crosslinking. Klemmt et al. [9] used agar and starch as a binder for wood fiber.

Sodium silicate can be used as a modulus-enhancing, hydrophobic additive in the processing of wood products. Organo-silanes react via condensation in a similar manner to sodium silicate, but with a slower reaction rate. Typically, methoxy, ethoxy, and propoxy groups replace the hydroxyl groups in sodium silicate. Kuan et al. [23] used tetramethoxysilane to crosslink wood flour in polyethylene composites. Crosslinking of polylactic acid with a methoxysilane through condensation reactions was earlier reported by Meekum and Kingchang [24]. The hydroxyl groups (-OH) can react with tetrafunctional or trifunctional silanes to form ether bonds, -C-O-C- with the release of water or alcohols. This condensation reaction can be either base- or acid-catalyzed relative to the isoelectric point of the organosilane. Sodium silicate and organosilanes react to form silica sand and either water or ethanol/methanol, so their use is environmentally friendly.

5.2 Fiber Orientation for Biodegradable 3DP

The orientation of fibers in materials such as paper and composites has a significant effect on the material's properties and structural performance, which has been extensively analyzed [25]. For example, Cox developed a mathematical model for continuous fibers [26], and Fu and Lauke developed a framework to calculate the structural performance of discontinuous fiber-reinforced polymers based on their fiber length distribution and the fiber orientation distribution, aimed at predicting the performance of injection molded parts [27].

In 3DP, fiber reinforcement has been used for a variety of processes and with a variety of fiber and binder types. The reinforcement is commonly used to increase tensile strength along the direction of the fibers. Namiki et al. used PLA thermoplastic extrusion as a basis for the integration of continuous carbon fiber reinforcement, which was mixed in the extruder prior to the extrusion [28]. Yang et al. developed a similar methodology based on fused deposition modeling (FDM) for continuous fiber reinforcement [29]. Both report improvements to tensile strength, but with reduced adhesion between fibers or layers. In contrast, Li et al. observed sufficient bonding at the nanoscale in a similar process [30]. Arunothayan et al. evaluated the behavior of steel fiber reinforcement for concrete 3DP and defined tensile strength based on nozzle design and fiber orientation. A smaller nozzle was found to increase fiber alignment and thereby the tensile strength along the extrusion direction [31]. Christ et al. [32] used fiber reinforcement for gypsum powder printing to achieve a significant increase in bending strength. They propose future work into biodegradable fibers. Balla et al. [33] utilized soy hull fiber with a thermoplastic binder for a fused filament fabrication method (FFFM).

5.3 Functional Grading/State of the Art for FGM in Wood

Functionally graded materials (FGM), often produced through functionally graded additive manufacturing (FGAM), are 3DP objects that consist of different material compositions, densities, or anisotropic fiber orientations at different locations in the object [34–36]. The material may vary continuously or discretely on a per voxel basis and may be designed to satisfy varying requirements at different part locations, such as different structural, thermal or other requirements. Ituarte et al. proposed the design and manufacturing of functionally graded materials at the macro-scale but has not implemented this design [37]. Duoss et al. developed 3DP foam-like structures at the micro-scale which exhibited different structural behaviors according to the placement of solid and void areas and the orientations of continuous fibers of material within an object [38]. Nakamura et al. proposed a methodology for the analysis of FGM at the micron-scale through micro-indentation [39].

Herrmann and Sobek propose the functional grading of concrete with varying densities by adding a lightweight filler that also provides thermal insulation. A spraying technique is used to create components that are graded according to structural and insulation requirements [40]. Craveiro et al. developed medium scale FGM 3DP using polyurethane polymers [41]. The work was developed further to include other materials, such as cement with a lightweight cork filler [42]. Torelli et al. analyzed different production methods for FGM and proposed a non-3DP method for horizontal or vertical layering. The FGM methods focused on isotropic material properties such as density and thermal performance but did not take anisotropic fibers into account [43]. Ahmed et al. developed a method for the functional grading of fiber reinforcement within concrete, whereby the directionality of the fibers was shown to have an impact on the structural performance of the 3DP pieces [44]. Lee et al. proposed a method of topology optimization (TO) to design FGM materials with different fiber orientations; however, the method does not take manufacturing constraints or toolpath design into account [45]. Hasanov et al. developed a small-scale method to produce FGM with fiber orientation along the direction of the extruded bead [46].

5.4 *Lignin in AM*

Lignin is a research material for AM. Useful lignin-based AM materials require extensive processing from black liquor produced in the chemical pulping process to produce a probably not biodegradable thermoplastic. Lignin is the second largest product of wood but is largely a waste product of the pulp and paper industry. Globally ~50 million tons of lignin are produced with most of it being used as a fuel source in paper mills. Due to its abundance and non-food source, lignin is an attractive biosourced material for AM. In the Kraft process white liquor (caustic aqueous solution of NaOH and Na_2S) is used to remove lignin from wood pulp producing black liquor which is a lignin rich waste stream. The caustic solution breaks up the ether linkages in lignin reducing the molecular weight and allowing it to be separated from cellulose and hemicellulose. Typically, black liquor is used directly as a fuel for the Kraft process. Due to the large quantities of lignin that are produced, there has been extensive interest in alternative, value added uses. Lignin typically has aromatic end-groups with an ether group, hydroxyl group, and unmodified methylenes. Chemical modification for improved AM use involves the addition of polar groups to the methylene or attaching ether or ester groups to the phenol groups. A wide range of reactions are possible. Addition of polar groups improves the solubility in water and processability for slurry-based AM. Attaching ether or ester groups enables targeted reactivity in crosslinking reactions for example in photo-induced crosslinking of an AM part [47].

5.5 *Nanocellulose in AM*

Nanocellulose (NC) is available in three types, nanocellulose fibers (CNF), nanocellulose crystals (CNC) and bacterial nanocellulose (BNC), Fig. 2. CNF or nanofibrillated cellulose (NFC) are produced by an enhanced enzymatic, chemical, and mechanical pulping process. CNFs are micron length fibers of 5 to 50 nm diameter used as rheology modifiers. CNC or nanowiskers are produced by acid hydrolysis of wood pulp and sonication. CNC are shorter fibers of 300 nm length and somewhat thinner diameter, 3–20 nm, compared to CNFs. CNCs are used in composites and coatings. BNCs are produced by bacteria through fermentation of sugar. BNCs are pure cellulose nanofibers of 20-100 nm diameter in a network. BNCs are used in filtration media and in some biomedical applications [48]. Figure 2 shows high resolution TEM images of CNCs and CNFs [49].

NC is usually processed as a high concentration gel (~90 % NC) using organic solvents or water followed by chemical, pH, or photo induced crosslinking to form a solid structure. Variation of the solvent composition is the primary tool to tune processing and properties. NC is biosourced, biodegradable and biocompatible. NC is expensive compared to other biodegradable nanomaterials, CNF ~$30/kg, CNC ~$300/kg, BNC ~$1,000/kg. Processing may have a significant environmental impact

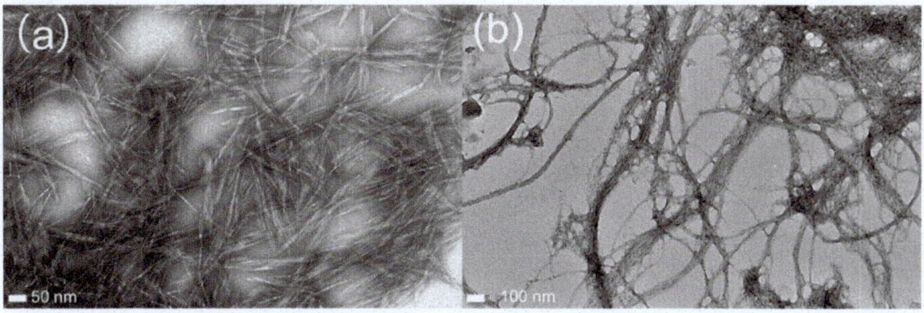

Fig. 2 Cellulose nano-crystals and cellulose nanofibers high resolution TEM from Xu et al. [49]

starting with wood pulping processes and ending with various chemical treatments usually in inefficient batch processes.

5.6 TEMPO oxidized nanocellulose

Nanocellulose can be improved in aqueous suspension stability, transparency and homogeneity, and mechanical properties especially in composites through oxidation using 2,2,6,6-tetramethylpiperidine-1-oxyl (TEMPO) and a co-oxidant (NaClO/NaBr). Alcohol groups in cellulose are converted into carboxylate groups (COO^-). This is a common modification for improved performance. The oxidation primarily impacts the surface charge on the nanofibers. TEMPO modification leads to increased hydrophilicity.

5.7 Acetylated Nanocellulose

Nanocellulose can be improved for organic suspension stability, barrier properties, and solubility in organic polymers especially in composites through acetylation of the hydroxyl groups using acetic anhydride. Alcohol groups in cellulose are converted into non-polar acetyl groups ($-OCOCH_3$). Acetyl modification leads to increased hydrophobicity. Variable degrees of acetylation are possible depending on the process and property needs.

Figure 3 shows examples of TEMPO and acetylation of nanocellulose fibers [50].

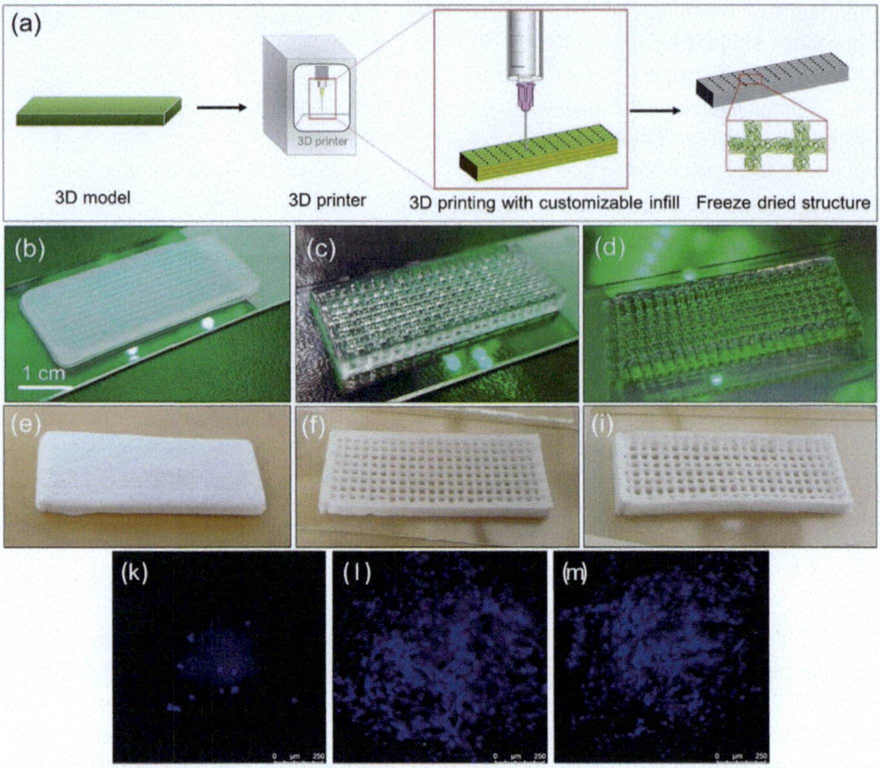

Fig. 3 **a** Direct ink writing using nanocelluloses. **b** CNF, **c** TEMPO modification (TOCNF), and **d** Acetylation modified (AceCNF). After freeze drying in **e**, **f**, and **i**. Objects loaded with algae in **k**, **l**, **m** at the bottom for CNF, TOCNF, and AceCNF grids. Reproduced from Adjary et al. [50] under CC-BY

6 Starch in AM

Starch is a biosourced biodegradable material that can be used in AM. Starch has been used extensively to 3D print food [51]. However, neat starch has poor mechanical properties in AM and for this reason requires processing additives. It is also commonly used as a binder especially in combination with cellulose as mentioned above [51].

7 Chitosan

Chitosan is a cationic linear polysaccharide composed of glucosamine units derived from chitin, a material from the exoskeleton of crustaceans through processing of chitin through deacetylation. Chitin is, for the most part, a byproduct of the seafood

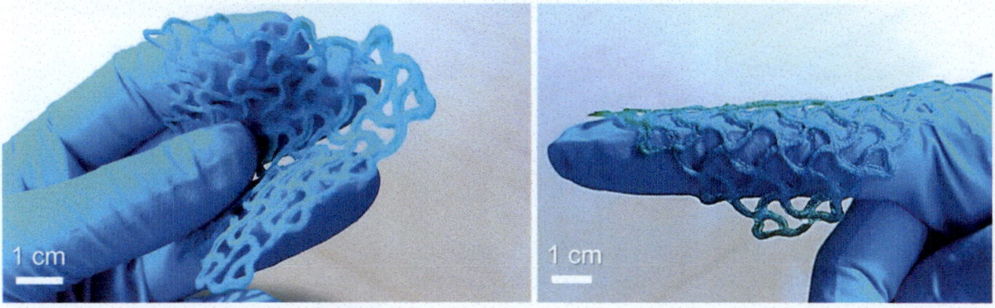

Fig. 4 Chitosan/CNF composite crosslinked using glutaraldehyde Reproduced from Ajdary et al. [54] under CC-BY-NC

industry, but it can also be made using fungi. Chitosan is biosourced and biodegradable as well as biocompatible for medical uses. The polymer forms an extrudable gel in water that can be crosslinked with methacrylic anhydride. Acrylic acid is used to control the viscosity of the sol/gel system. It can also be blended with biodegradable or biosource polymers for an extrudable composite resin. Chitosan is priced like PHB, about $15/kg, therefore about 5 times the price of PLA and 8 times the price of starch, limiting its widespread use to specialty applications. It is a main polymer used in bioprinting for medical implants and drug delivery [52, 53]. Figure 4 shows an example of a chitosan/CNF composite made by 3D printing [54].

The combination of cellulose fiber and chitin can produce an AM resin with properties like polyurethane foam that can be worked like wood [55]. There is a biological model for this mixture of cellulose and chitin in oomycetes which are essentially fungus-like algae. This extrudable material with low shrinkage is called fungus-like additive material (FLAM). Figure 5 shows some of the features of FLAM [55].

8 Engineered Living Materials

There is current interest in the incorporation of living organisms in functional materials termed *animate materials* [52]. The intent is to provide biosourced, biodegradable, functional materials that take advantage of the properties of living matter such as self-healing, growth, oxygen production, CO_2 consumption, self-optimizing depending on environmental conditions, self-cleaning and self-maintaining. Some examples in AM include fungi, fungus-like additive material (FLAM), algae, and bacteria. The field of engineered living materials overlaps with biomedical implant and medical device AM.

Fig. 5 3D printed FLAM composite material showing processability. Reproduced from Sanandiya et al. [55] under CC-BY

8.1 Fungi in AM

Fungi offer the potential for 3D printing of regenerative, environmentally adaptive materials with growth, communication and decision-making properties. For example, the Swiss/Dutch group of Gantenbein et al. [56] has fabricated robust robotic skins

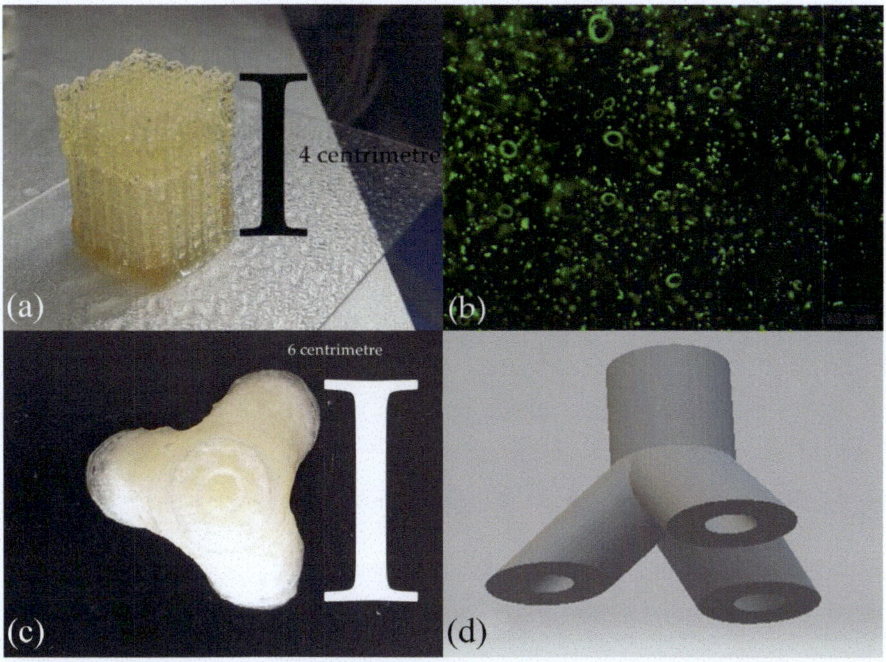

Fig. 6 **a** 3D-printed 113 layer structure of gelatin/alginate/agarose and structure loaded with **b** adipose-derived stem cells. **c** and **d** show a trigeminal vasculature structure intended as a model implant structure. Reproduced under CC BY 4.0 from Gong et al. [57]

by 3D printing agar gels containing nutrients and inoculating fungi which later fill air gaps in the designed skin with a tough, hydrophobic, fibrous mycelium network that is self-healing in some circumstances. The growth of the mycelium can be controlled initially though nutrient concentration which results in controlled fibrous density and skin toughness. The coating is biodegradable.

Gantenbein et al. [56] 3D printed agar gel laden with fungus to produce living skin for a robotic sphere that could clean up spills. They point out that the exploration and exploitation modes of fungal growth could be manipulated by controlling the malt food supply.

Figure 6 is an example of 3D printing of animate materials from Gong et al. [57].

8.2 Algae in AM

Algae produce anionic polysaccharides that can form hydrogels either naturally or with cationic crosslinking salts such as Ca^{+2}. 3D soft gels can be printed for applications such as drug delivery. Examples are alginate, agarose, carrageenan, fucoidan.

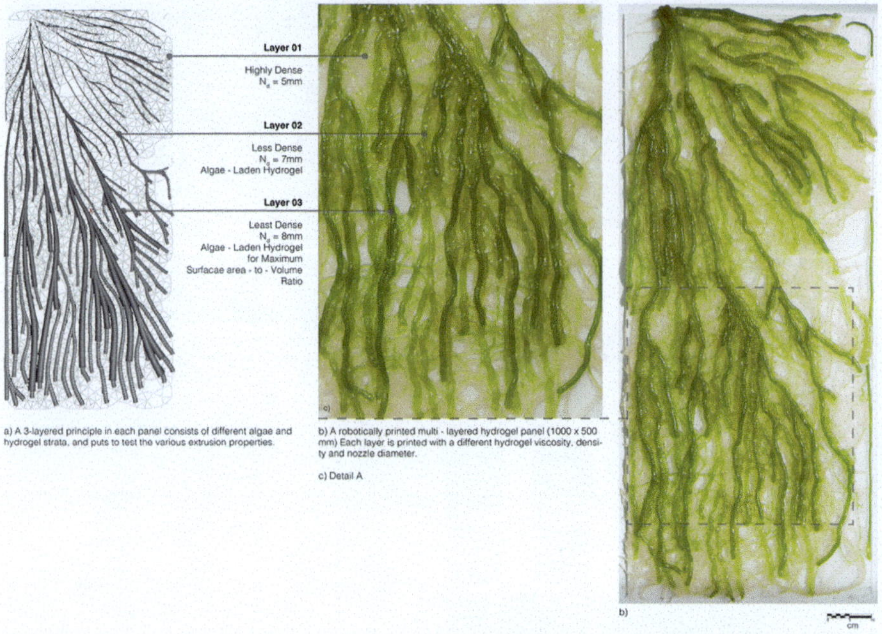

Fig. 7 A 3D printed gradient of algae loading to produce leaf-like structure. Reproduced from [59] under CC-BY

AM using Algae sourced polyscharides is a research topic with no current applications. Algae can also be used to dope generally non-biodegradable hydrogels to produce living 3D printed sheets for removal of metals from water and to optimize algae cell culture. [58]

Figure 7 shows an example of large-scale AM of leaf structures using hydrogel printing [59].

8.3 *Bacteria in AM*

AM using bacteria is of interest for retention of bacteria in bioreactors, creation of controlled biofilms for water treatment, and in medical devices under study. Generally, bacteria are carried by a hydrogel which can be petroleum based like polyethylene oxide, Pluronics or polyvinyl alcohol, or biobased polysaccharides like sodium alginate, hyaluronic acid, agarose (agar), or chitosan (chitin), or biobased proteins like collagen, fibrin, or gelatin. Petrochemical based hydrogels are not biodegradable. Microbial co-cultures can be created using layer-by-layer 3D printing and these can be symbiotically designed.

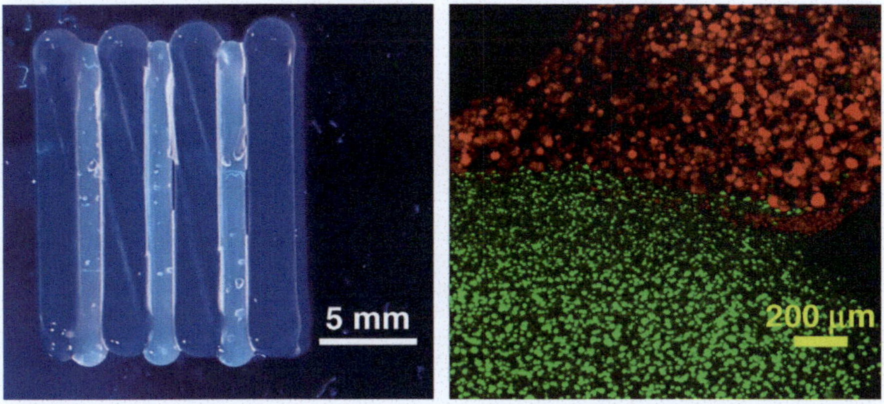

Fig. 8 3D printed dual cell structure of yeast and bacteria (normally incompatible) Left: a UV-illuminated printed stripes containing RFP yeast and GFP bacteria. Right: confocal microscopy of the interface of the stripes. Reproduced from Johnston et al. under CC-BY [60]

9 Composite Materials

Biobased and synthetic biodegradable materials are often blended to improve processing, sensitivity to water, and mechanical properties or to induce functionality such as in Algae-PLA composites. PLA and PHAs are often blended in immiscible blends to improve on the mechanical properties of PHAs and to improve the compostability of PLA. Commercial blends of PLA with wood fiber are common for melt extrusion fibers. TPS is often blended with PLA or PHAs to reduce the hydrophilicity and to improve mechanical properties.

Figure 8 demonstrates how AM can be used to produce new biologically active composite materials [60].

10 Summary of Biodegradable Materials in AM

A wide range of materials are used to produce biodegradable 3D printing inks for AM. Materials range from bacteria-laden hydrogels to thermoplastic resins. In this complex landscape it is important to consider the lack of real compostability in the US for PLA, and the potential production of microplastic waste during degradation in blends of truly compostable materials and essentially non-compostable materials for example, greenwashing of petrochemical based plastics by the addition of starch resulting in microplastics waste.

Choice of biodegradable inks depends largely on the cost, intended processing and end-use. For many applications PHAs are an ideal material but the cost excludes their use from many applications. Wood and wood composites could be useful for large-scale applications but cannot, so far, generate sufficient structural precision for many applications without further working of the printed part. Most adaptive/living biodegradable materials are based on doped hydrogels.

References

1. Kokare, S., Oliveira, J.P., Godina, R.: Life cycle assessment of additive manufacturing processes: a review. J. Manuf. Sys. **68**, 536–559 (2023)
2. Biermann, C.J.: Handbook of Pulping and Papermaking, 2nd edn. Elsevier, New York (1996)
3. Le Duigou, A., Castro, M., Bevan, R., Martin, N.: 3D printing of wood fibre biocomposites: from mechanical to actuation functionality. Mater. Des. **96**, 106–114 (2016)
4. Wang, Q., Sun, J., Yao, Q., Ji, C., Liu, J., Zhu, Q.: 3D printing with cellulose materials. Cellulose **25**, 4275–4301 (2018)
5. Davide, S.: UPM demonstrates large scale robotic 3D printing of wood polymer composites (2019). Last accessed on 10 May 2020. https://www.3dprintingmedia.network/upm-demonstrates-large-scale-robotic-3d-printing-of-wood-based-polymer-composites/
6. Kariz, M., Sernek, M., Kitek-Kuzman M.: Use of wood powder and adhesive as a mixture for 3D printing. Eur. J. Wood Prod. **74**, 123–126 (2016)
7. Rosenthal, M., Henneberger, C., Gutkes, A., Bues, C.T.: Liquid deposition modeling: a promising approach for 3D printing of wood. Euro. J. Wood Wood Prod. **76**, 797–799 (2018)
8. Kam, D., Layani, M., Minerbi, S.B., Orbaum, D., Ben-Harush, S.A., Shoseyov, O., Magdassi, S.: Additive manufacturing of 3D structures composed of wood materials. Adv. Mater. Technol. **4**, 1900158 (2019). https://doi.org/10.1002/admt.201900158
9. Klemmt, C., Meibodi, M.A., Beaucage, G., Mcgee, W.: Large-scale robotic 3D printing of plant fibre and bioplastic composites co-creating the future—eCAADe 40 **1**, 9–18 (2022) and Klemmt, C., Beaucage, G.: Accounting for expansion and shrinkage in fiber-reinforced composite extrusion of wood-based composites submitted to 3D printing and adding manufacturing (2025) and Chapter 3.3 in this book
10. Henke, K., Treml, S.: Wood based bulk material in 3D printing processes for applications in construction. Eur. J. Wood Prod. **71**, 139–141 (2013)
11. Zeidler, H., Klemm, D., Böttger-Hiller, F., Fritsch, S.: 3D printing of biodegradable parts using renewable biobased materials. Procedia Man. **21**, 117–124 (2018)
12. Hausmann, M.K., Siqueira, G., Libanori, R., Kokkinis, D., Neels, A., Zimmermann, T., Studart, A.R.: Complex-shaped cellulose composites made by wet densification of 3D printed scaffolds. Adv. Funct. Mater. **30**, 201904127 (2019). https://doi.org/10.1002/adfm.201904127
13. Deville, A., Saiz, E., Nalla, R.K., Tomsia, A.P.: Freezing as a path to build complex composites. Science **311**, 515–518 (2006)
14. Jiang, B., Yao, Y., Liang, Z., Gao, J., Chen, G., Xia, Q., Mi, R., Jiao, M., Wang, X., Hu, L.: Lignin-based direct ink printed structural scaffolds. Small **16**, 201907212 (2020). https://doi.org/10.1002/smll.201907212
15. Ansell, M.P.: Wood Composites. Elsevier, New York (2015). https://doi.org/10.1016/C2014-0-02700-1
16. Theagarajan, R., Moses, J.A., Anandharamakrishnan C.: 3D Extrusion printability of rice starch and optimization of process variables. Food Bioprocess Tech. **13**, 1048–1062 (2020). https://doi.org/10.1007/s11947-020-02453-6
17. Chen, H., Xie, F., Chen, L., Zheng, B.: Effect of rheological properties of potato, rice and corn starches on their hot-extrusion 3D printing behaviors. J. Food Eng. **244**, 150–158 (2019)
18. Tian, P., Yang, F., Xu, Y., Lin, M.M., Yu, L.P., Lin, W., Lin, Q.F., Lu, Z.F., Huang, S.Y., Chen, Y.Z.: Oral disintegrating patient-tailored tablets of warfarin sodium produced by 3D printing. Chen. Drug Dev. Ind. Pharm. **44**, 1918–1923 (2018)
19. Yang, L., Tang, S., Li, G., Qian, L., Mei, J., Jiang, W., Fan, Z.: Layered extrusion forming of complex ceramic structures using starch as T removable support. Cer. Int. **45**, 21843–21850 (2019)

20. Holland, S., Tuck, C., Foster, T.: Fluid gels: a new feedstock for high viscosity jetting. Food Biophys. **13**, 175–185 (2018). https://doi.org/10.1007/s11483-018-9523-x
21. ETHZ print bacteria colony (2020). https://www.youtube.com/watch?v=W2piUVYukyA
22. Wang, J., Liu, Y., Su, S., Wei, J., Rahman, S.E., Ning, F., Christopher, G., Cong, W., Qiu, J.: Ultrasensitive wearable strain sensors of 3D printing tough and conductive hydrogels. Polymers **11**, 1873–1887 (2019). https://doi.org/10.3390/polym11111873
23. Kuan, C.F., Kuan, H.C., Ma, C.C.M., Huang, C.M.: Mechanical, thermal and morphological properties of water-crosslinked wood flour reinforced linear low-density polyethylene composites. Compos.: Part A **37**, 1696–1707 (2006)
24. Meekum, U., Kingchang, P.: Peroxide/Silane crosslinked Poly(lactic acid) hybrid biocomposite reinforced with empty fruit bunch and cotton fibers for hot-fill food packaging. BioResources **12**, 5086–5101 (2017)
25. Hyer, M.W., White, S.R.: Stress Analysis of Fiber-Reinforced Composite Materials. DEStech Publications Inc, Lancaster, PA USA (2009)
26. Cox, H.L.: The elasticity and strength of paper and other fibrous materials. Brit. J. Appl. Phys. **3**, 72–79 (1952)
27. Fu, S.Y., Lauke, B.: Effects of fiber length and fiber orientation distributions on the tensile strength of short-fiber-reinforced polymers. Comp. Sci. Tech. **56**, 1179–1190 (1996)
28. Namiki, M., Ueda, M., Todoroki, A., Hirano, Y., Matsuzaki, R.: 3D printing of continuous fiber reinforced plastic. In: SAMPE Seattle 2014 International Conference and Exhibition. International SAMPE Technical Conference, Social for the Advancement of Material and Process Engineering, SAMPE Tech Seattle 2014 Conference, Seattle, United States (2014)
29. Yang, C., Tian, X., Liu, T., Cao, Y., Li, D.: 3D printing for continuous fiber reinforced thermoplastic composites: mechanism and performance. Rap. Proto. J. (2017)
30. Li, N., Li, Y., Liu, S.: Rapid prototyping of continuous carbon fiber reinforced polylactic acid composites by 3D printing. J. Mat. Proc. Tech. **238**, 218–225 (2016)
31. Arunothayan, A.R., Nematollahi, B., Ranade, R., Bong, S.H., Sanjayan, J.G., Khayat, K.H.: Fiber orientation effects on ultra-high performance concrete formed by 3D printing. Cem. Conc. Res. **143**, 106384 (2021)
32. Christ, S., Schnabel, M., Vorndran, E., Groll, J., Gbureck, U.: Fiber reinforcement during 3D printing. Mat. Let. **139**, 165–168 (2015)
33. Balla, V.K., Tadimeti, J.G.D., Kate, K.H., Satyavolu, J.: 3D printing of modified soybean hull fiber/polymer composites. Mat. Chem. Phys. **254**, 123452 (2020)
34. Loh, G.H., Pei, E.: Building a conceptual understanding of functionally graded additive manufacturing (2019). https://bura.brunel.ac.uk/bitstream/2438/18693/1/FullText.pdf. Last accessed 25 Feb 2024
35. Elishakoff, I., Pentaras, D., Gentilini, C.: Mechanics of Functionally Graded Material Structures. World Scientific Publishing, Hakensack NJ (2016)
36. Mahamood, R.M., Akinlabi, E.T.: Functionally Graded Materials. Springer Publishing AG, Cham Switzerland (2017)
37. Ituarte, I.F., Boddeti, N., Hassani, V., Dunn, M.L., Rosen, D.W.: Design and additive manufacture of functionally graded structures based on digital materials. Add. Manuf. **30**, 100839 (2019)
38. Duoss, E.B., Weisgraber, T.H., Hearon, K., Zhu, C., Small, I.V., Metz, T.R., Vericella, J.J., Barth, H.D., Kuntz, J.D., Maxwell, R.S., Spadaccini, C.M.: Three-dimensional printing of elastomeric, cellular architectures with negative stiffness. Adv. Fun. Mat. **24**, 4905–4913 (2014)
39. Nakamura, T., Wang, T., Sampath, S.: Determination of properties of graded materials by inverse analysis and instrumented indentation. Acta Mat. **48**, 4293–4306 (2000)
40. Herrmann, M., Sobek, W.: Functionally graded concrete: numerical design methods and experimental tests of mass-optimized structural components. Struc. Conc. **18**, 54–66 (2017)
41. Craveiro, F., Bártolo, H., Bártolo, P.J.: Functionally graded structures through building manufacturing. Adv. Mat. Res. **683**, 775–778 (2013)
42. Craveiro, F., Nazarian, S., Bartolo, H., Bartolo, P.J., Duarte, J.P.: An automated system for 3D printing functionally graded concrete-based materials. Add. Manuf. **33**, 101146 (2020)

43. Torelli, G., Fernández, M.G., Lees, J.M.: Functionally graded concrete: design objectives, production techniques and analysis methods for layered and continuously graded elements. Cons. Build. Mat. **242**, 118040 (2020)
44. Ahmed, Z.Y., Bos, F.P., van Brunschot, M.C.A.J., Salet, T.A.: On-demand additive manufacturing of functionally graded concrete. Virt. Phys. Proto. **15**, 194–210 (2020)
45. Lee, J., Kim, D., Nomura, T., Dede, E.M., Yoo, J.: Topology optimization for continuous and discrete orientation design of functionally graded fiber-reinforced composite structures. Comp. Struct. **201**, 217–233 (2018)
46. Hasanov, S., Gupta, A., Nasirov, A., Fidan, I.: Mechanical characterization of functionally graded materials produced by the fused filament fabrication process. J. Manuf. Proc. **58**, 923–935 (2020)
47. Wu, T., Sugiarto, S., Yang, R., Sathasivam, T., Weerasinghe, U.A., Chee, P.L., Yap, O., Gustav Nystro, G., Kai, D.: From 3D to 4D printing of lignin towards green materials and sustainable manufacturing. Mater. Horiz. **12**, 2789–2819 (2025)
48. Ee, L.Y., Li, S.F.Y.: Recent advances in 3D printing of nanocellulose: structure, preparation, and application prospects Nanoscale Adv. **3**, 1167–1208 (2021)
49. Xu, X., Liu, F., Jiang, L., Zhu, J.Y., Haagenson, D., Wisenborn, D.P.: Cellulose nanocrystals versus cellulose nano brils: a comparative study on their microstructures and eects as polymer reinforcing agents. ACS Appl. Mater. Interfaces **5**, 2999–3009 (2013)
50. Ajdary, R., Huan, S., Ezazi, N.Z., Xiang, W., Grande, R., Santos, H.A., Rojas, O.J.: Acetylated nanocellulose for single-component bioinks and cell. Proliferation on 3D-printed scafolds. Biomacromolecules **20**, 2770–2778 (2019)
51. Ronga, L., Chen, X., Shen, M., Yanga, J., Qia, X., Li, Y., Xie, J.: The application of 3D printing technology on starch-based product: a review. Trnd. Food Sci. Tech. **134**, 149–161 (2023)
52. Ball, P.: Animat. Mater. MRS Bull. **46**, 553–559 (2021)
53. Taghizadeh, M., Taghizadeh, A., Yazdi, M.K., Zarrinta, P., Stadler, F.J., Ramsey, J.D., Habibzadeh, S., Rad, S.H., Naderi, G., Saeb, M.R., Mozafari, M., Schubert, U.S.: Chitosan-based inks for 3D printing and bioprinting. Green Chem. **24**, 62–101 (2022)
54. Ajdary, R., Reyes, G., Kuula, J., Raussi-Lehto, E., Mikkola, T.S., Kankuri, E., Rojas, O.J.: Direct ink writing of biocompatible nanocellulose and chitosan hydrogels for implant mesh matrices. ACS Polym. Au **2**, 97–107 (2022)
55. Sanandiya, N.D., Vijay, Y., Dimopoulou, M., Dritsas, S., Fernandez, J.G.: Large-scale additive manufacturing with bioinspired cellulosic materials. Sci. Rep. **8**, 8642 (2018)
56. Gantenbein, S., Colucci, E., Käch, J., Trachsel, E., Coulter, F.B., Rühs, P.A., Masania, K.: Studart AR three-dimensional printing of mycelium hydrogels into living complex materials. Nat. Mat. **22**, 128–134 (2023)
57. Gong, C., Kong, Z., Wang, X.: The effect of agarose on 3D bioprinting. Polymers **13**, 4028 (2021)
58. Grira, S., Khalifeh, H.A., Alkhedher, M., Ramadan, M.: 3D printing algae-based materials: pathway towards 4D bioprinting. Bioprinting **33**, e00291 (2023)
59. Malik, S., Hagopian, J., Mohite, S., Lintong, C., Stoffels, L., Giannakopoulos, S., Beckett, R., Leung, C., Ruiz, J., Cruz, M., Parker, B.: Robotic extrusion of Algae-Laden hydrogels for large-scale applications. Glob. Chall. **4**, 1900064 (2020)
60. Johnston, T.G., Yuan, S.F., Wagner, J.M., Yi, X., Saha, A., Smith, P., Nelson, A., Alper, H.S.: Compartmentalized microbes and co-cultures in hydrogels for on-demand bioproduction and preservation. Nat. Comm. **11**, 563 (2020)

Gregory Beaucage is Professor of Chemical and Materials Engineering at the University of Cincinnati. He has published extensively in polymers and materials science with 189 publications and 6,900 citations. He has collaborated with Christoph Klemmt on the development of new materials for 3D printing of wood for the past four years. Beaucage holds a Ph.D. from the University of Massachusetts at Amherst in Polymer Science and Engineering and was a post doc at

Sandia National Laboratories before joining the faculty at the University of Cincinnati. He was a Fulbright Global Scholar at Sheffield University in the UK and at several Universities in Ethiopia, South Africa, and Lesotho. He has been Visiting Professor at University of the Witwatersrand, Johannesburg, University of Sheffield, University of Cape Town, and ETH in Zürich.

Christoph Klemmt graduated from the Architectural Association in London and received his Doctor Technicae from the University of Applied Arts Vienna. He has worked amongst others for Zaha Hadid Architects, where his responsibilities were focused on the company's projects in China. He is Associate Professor at the Department of Design, Art, Architecture and Planning at the University of Cincinnati, where he set up and runs the Architectural Robotics Lab. He co-founded Orproject, an architect's office specializing in advanced geometries with an ecologic agenda. Orproject has been featured widely and has exhibited at the Palais De Tokyo in Paris, the China National Museum in Beijing and the Biennale in Venice.

Part II
Materials and Methods

Chapter 4
FLAM: 3D Printing with Chitin—Cellulose Composites

Stylianos Dritsas and Javier G. Fernandez

Abstract The use of natural biological composite materials offers a unique opportunity to improve our sustainability patterns of production and consumption because of their abundant availability in nearly every ecosystem, their low energy manufacturing requirements and eventually their natural biodegradation. This chapter presents the design of the Fungus-Like Adhesive Materials (FLAM), a family of cellulose-chitin biopolymers, and the development of its robotic 3D printing technology. The material's bio-inspired design, sustainability consideration and its characteristics are documented aiming to stress the pitfalls of modern bioplastics and propose an alternative approach to sustainable materials. Furthermore, the 3D printing technology's development highlights the unique challenges and considerations required for enabling the use of water-based biological materials. The process characteristics, including their differences when compared to synthetic polymers, challenges in controlling their shrinkage during curing as well as the opportunities to deploy contemporary process control methodology, such as machine learning to address their natural variability are shared. Several experiments are presented aiming to demonstrate the versatility of FLAM including demonstrators of a wide range of scales. Additionally, their remarkable ability to be synthesized from urban waste sources, namely paper and food waste, highlights their potential for closed-loop urban metabolic cycles. Finally, several potential future applications of FLAM are presented. Those aim toward use-cases where the technology is deployed in production where cellulosic materials are already used, for substituting existing processes and products involving synthetic polymers, contributing toward the problem of plastics, as well as for scenarios where a new material process may seed design innovation.

Keywords Biomaterials · Cellulose · Chitin · 3D printing · Sustainability · FLAM

S. Dritsas (✉)
Architecture and Sustainable Design, SUTD, Singapore, Singapore
e-mail: dritsas@sutd.edu.sg

J. G. Fernandez
Engineering Product Development, SUTD, Singapore, Singapore

© The Author(s), under exclusive license to Springer Nature Singapore Pte Ltd. 2026

C. Klemmt and G. Beaucage (eds.), *Biodegradable 3D-Printing*, The Materials Research Society Series, https://doi.org/10.1007/978-981-95-4025-9_4

1 Introduction

In nature, "manufacturing" takes place under very different circumstances compared to those involved in the industrial production of man-made goods. Material components are sourced locally, without requiring transcontinental extraction and shipping. Additionally, assembly takes place with miniscule requirements of energy, unlike the vast resources consumed by large-scale industrial processes. Most importantly, however, at the end of their life, natural forms are integrally reclaimed within their environment, unlike industrial products which require human intervention to manage streams of waste. Research in manufacturing using natural composite materials was motivated by the understanding that the modern historic paradigms of production and consumption may need to be fundamentally revisited given their environmental repercussions. The approach was to replicate nature's manufacturing approach within today's technological capability landscape. Self-imposed constraints include: the materials employed had to be widely available and at large quantities and low cost, their assembly had to be frugal in the consumption of material resources and energy, and the products had to retain their natural characteristics to ensure that they can be naturally reclaimed at the end of their life without human intervention.

Cellulose is the most abundant polymer on Earth most associated with plants, and it is extensively used in paper and textile products [1–3]. Chitin is the second most widely available [4–6] but lesser-known biopolymer encountered in a wide range of organisms including arthropods, insects, and mushrooms, and it is used for agricultural, industrial, and biomedical applications. The natural production of these two polysaccharides is in billions of tons annually [2, 4]. For its abundance and renewability, cellulose and its derivatives played a crucial role in the early development of biopolymer manufacturing applications, leading to some of the first plastics before the invention of synthetic polymers like nylon. This includes thermoplastics such as celluloid films and cellophane, and fibers like rayon [7]. The invention of synthetic plastics led to their market dominance and halted bioplastic development, establishing a manufacturing paradigm centered on synthetic polymers. Interest in bioplastics resurfaced only after the environmental impact of synthetics became apparent [8]. However, the industry's decades of expertise and substantial investment in equipment for petroleum-derived polymers drive a desire for bioplastics that are immediately usable in existing manufacturing processes like injection molding and 3D printing. Prioritizing compatibility with established synthetic polymer techniques over environmental benefits requires significant modifications to the bioplastics, which harm their unassisted integration within ecological processes and often involve the use of strong organic solvents and hazardous pollutants such as acetone, carbon disulfide, and sulfuric acid [9]. Another pitfall to be avoided is found in modern bioplastics such as polylactic acid (PLA), which is produced from plant matter such as corn and sugarcane. While its components originate in renewable sources, its production requires the allocation of agricultural land and the use of food sources, and, at the end of their life PLA products require composting in specialized facilities [10, 11].

2 Material Design

Fungus-Like Adhesive Materials (FLAM) are a family of cellulosic-chitinous biopolymers whose development was informed by the above considerations. Apart from cellulose and chitin, the other two components in FLAM are water (the solvent) and acetic acid, with same concentration as common table vinegar. The acetic acid is used to introduce positive charges on the functional groups of chitosan, forcing the long polymeric chains into dissolution [12]. The term fungus-like is associated with oomycotes, a microorganism historically misclassified as fungi, which are some of the few encountered in nature that combine cellulose and chitin in their cell walls [13, 14]. FLAM's development is the result of earlier research into chitinous biopolymers like Shrilk—an insect cuticle-mimicking material composed of fibroin (the protein from spider silk) and chitin extracted from shrimp shells [15]. One of the challenges with chitinous biopolymers that FLAM sought to address was creating materials capable of forming 3D objects. While chitinous polymers are suitable for producing 2D films and laminates, they struggle to maintain a 3D shape due to their initial liquid state. Additionally, the shrinkage forces exerted during the drying process of chitinous films require a secondary material component to assist in retaining the desired shape. Cellulose is highly compatible with chitin, with which it shares its structural backbone, and capable of forming homogenous composites and it was selected exactly for this purpose. Nevertheless, chitin may also be used to bond inorganic materials such as minerals to create bio mortars [16] as well as metal particles to create conductive coatings [17, 18]. Therefore, chitin acts as a binder and cellulose as a filler or fibrous reinforcement in a conventional composite materials sense. FLAM were developed for additive manufacturing aiming to achieve spatially localized material assembly, in the sense of deploying material only where needed. This was aimed at on the one hand avoiding subtractive manufacturing, where substantial waste is produced by removal from a uniform material stock to achieve a desired shape, and on the other to investigate the opportunities of non-uniform material patterning. Selective spatial patterning of materials is a key principle of hierarchical design encounter in natural organisms [19] which may be approached by additive manufacturing processes.

FLAMs can be produced from either industrial grade pure cellulose or wood fiber waste from timber and paper manufacturing. The density of dry FLAMs' is 0.37 g/cc and Young's modulus of 0.26 GPa are in the range of high-density synthetic foams, for example machinable polyurethanes, and low-density natural timbers, such as balsa and cedar (Fig. 1a). Using pure cellulose fibers produces materials with paper-like appearance, while using wood flour produces FLAM that resembles low-density particulate composites (Fig. 1b, c). Alternative sources of cellulose fiber investigated include waste natural and engineered wood, paper products, and cotton textiles from disposed garments. Pure cellulose produces the strongest composites because contamination such as adhesives and inks in plywood and paper sources, hemicellulose, lignin and oils in wood waste as well as chemical treatments used for color retention in textiles are detrimental in the resultant FLAMs' mechanical properties. The chitosan used is sourced as a by-product of the shrimp farming

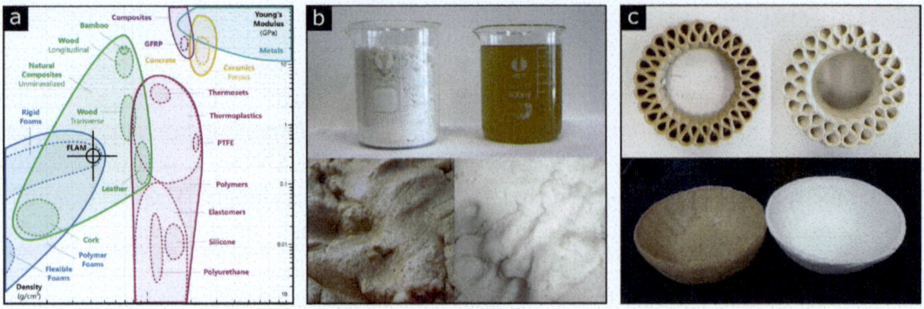

Fig. 1 **a** Ashby plot of density and Young's modulus of materials with relevant properties to FLAM. **b** Cellulose and chitosan and FLAM with wood flour and pure cellulose. **c** 3D printed and cast FLAM with wood flour and pure cellulose

industry, for agricultural use, which has significantly lower cost than higher purity chitosan for medical and cosmetic applications. The approximate cost of FLAM is 2USD/Kg which is competitive to synthetic petrochemical polymers. Additional properties include thermal conductivity, electrical conductivity and flammability, which resemble those of cellulosic materials and are comparable with synthetic polymers. Most importantly, because the synthesis of the biopolymer composite does not alter the source materials in an irreversible manner, FLAM retains their biodegradability of cellulose and chitin [20].

3 Digital Manufacturing

FLAM's rheologic properties were originally tuned for additive manufacturing, namely extrudability and shape retention [21, 22]. These are associated with the water-to-dry content ratio as well as particle size, subject to 0.2 MPa supply pressure constraints. Higher pressures result in phase separation where the dissolved chitin extrudes without the cellulose. While increased water content eases extrudability, shape retention becomes a problem as the material slumps under self-weight. Additionally, large particle sizes ease shape retention but require higher extrusion pressures. Moreover, the ratio of chitin is dictated by the solubility of chitosan and affects both extrudability and shape retention. Too low, by weight ratio (<1%) results in brittle objects but too high (>5%) is also undesirable because objects deform or tear under the shrinkage forces during drying. Therefore, experimental design resulted in the properties of wet FLAM, namely 1:8 chitin to cellulose by weight in 3% chitosan solution, using cellulose with 100–500 μm fiber length results in bulk density of 1.15 g/cc and dynamic viscosity in the range of 1.0 to 1.5 kPa s.

The extrusion system created for FLAM is based on an industrial articulated arm which serves as the positioning system paired with an industrial bulk material unloading and dispensing system commonly used for high viscosity applications

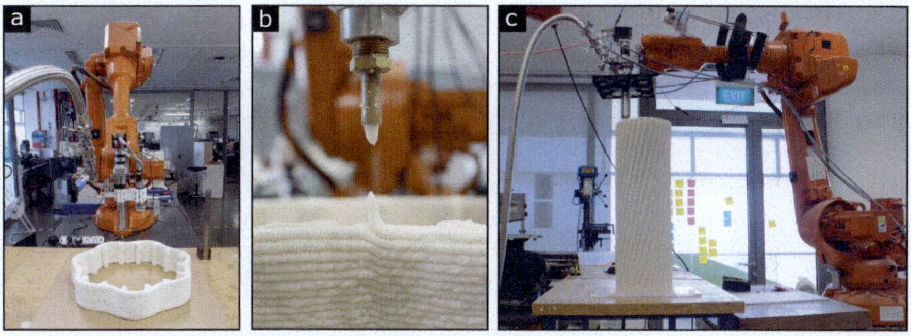

Fig. 2 **a** Industrial robotic 3D printing setup. **b** Detail of extrusion using 7 mm nozzle diameter. **c** Rapid prototyping setup with 1.5 mm nozzle diameter and in-process drying fans

such as adhesives and sealants. The extrusion rate of FLAM depends on the positioner's velocity and diameter of the nozzle used. Typical settings of 50 mm/sec linear velocity and 115 cc/min flow rate at 7 mm and 36 mm/sec, 3.8 cc/min with 1.5 mm nozzle diameter. Extrusion settings were derived experimentally using the Response Surface Modeling (RSM) methodology [23]. Linear velocity, material flow rate, and layer height were the input parameters of the model and the extruded FLAM's dimensions, and tensile strength were its outputs. A quadratic model was fitted to the collected observations and the desired process settings were derived using numerical optimization finding the process settings that could achieve the targeted dimensions. An interesting aspect of the RSM model is that for the same targeted dimensions it produces more than one solution. The difference between those is the tensile strength of the 3D printed material. Therefore, it is possible for objects of the same shape to exhibit different mechanical properties and even achieve variation within the same object by continuous modulation of process settings. Additionally, the ability to extrude FLAM at even at 100 mm/sec with 7 mm nozzle diameter, for objects with curvature continuity, is between one and two orders of magnitude higher than thermoplastic 3D printers. This is because the material is deposited at room temperature therefore extrusion speed is not constrained by the melting temperature of the polymer at the hot end. This is what slows down thermoplastic extrusion as nozzle diameter increases. Instead, the maximum positioner's speed and accuracy and the extruders flow rate are the main constraints in FLAM 3D printing (see Fig. 2).

4 Process Characteristics

After FLAM objects are 3D printed, they undergo curing by evaporative exchange of their water content with air. During the drying process, capillary forces exert internal stresses causing undesirable dimensional changes. Change of mass due to loss of water is substantial with dry-to-wet ratio of 0.245 (0.007), that is nearly

75% of the composite's mass evaporates. The associated dimensional changes of the objects, or shrinkage, exhibit anisotropic characteristics which may be attributed to several factors. The extrusion process itself is a key contributor to anisotropy because material continuity varies among all principal directions. The shrinkage along the extrusion direction is 0.05 (0.014), transversely is 0.173 (0.013) and vertically is 0.289 (0.04). From those, vertical shrinkage is quite substantial as objects are reduced to nearly two thirds of their original height (Fig. 3a). Vertical shrinkage exhibits a linear relationship with height and can be mitigated by printing additional layers from a scaled version of the design geometry (Fig. 3b). This systemic behavior offers an opportunity for taking advantage of this otherwise undesirable side effect, to achieve improved resolution after the parts have cured as their layers become finer. Additionally, boundary conditions also affect drying such as the first layer's ability to dehydrate and shrink is reduced by the adhesion to the base plate while the topmost layer has higher surface area ratio exposed to air and thus dries faster. Curing also takes place top-down and inside-out due to gravity and moisture gradients downwards and toward the exterior surface, respectively. The design geometry is also contributing to drying time and shrinkage. Completely solid objects or with thick cross-sectional profiles dry slower than objects with thinner profiles, internal cavities or in general larger surface areas to volume ratios.

The environmental conditions in terms of temperature and humidity and more broadly the drying regime also affect the rate of curing and the potential for shape deformation (Fig. 3c). These are not only critical in the sense of geometric accuracy but also production speed from design to manufacturing finished objects. Indicatively, at room temperature of 24 °C and 75% relative humidity (RH) specimens with 10 mm wall thickness required approximately ten days before reaching constant mass. However, using forced air convection at 27 °C and 80% RH reduces the same object's drying time to slightly over two days. Using hot air at 50 °C and 42% RH further reduces drying time below 24 hours without difference in shape and mass. While drying can be further reduced there are constraints in the highest temperature, namely circa 60 °C, where the organic material components begin to deteriorate. Experiments in exposing wet FLAM objects to sudden low humidity conditions, namely using a dehumidifier within an enclosure at 20% RH, caused unpredictable

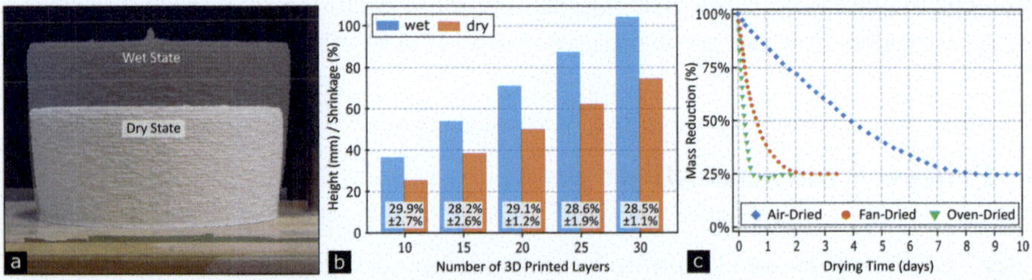

Fig. 3 **a** Photograph overlay of the same wet and dry 20-layer/100 mm diameter 3D printed cylinder. **b** Vertical shrinkage is proportional to the number of layers 3D printed. **c** Drying time profiles under different temperature and relative humidity regimes

deformations and layer delamination. Nevertheless, in small-scale rapid prototyping experiments, with 1.5 mm nozzle diameter, at room temperature conditions, using forced air convection by attaching fans to the extruder, it was possible to dry in-process the material to enable 3D printing hundreds of layers vertically without buckling under self-weight [24]. Overall, unlike thermoplastics where temperature and pressure are the two key in-process parameters for control, biomaterials such as FLAM require post-process water content control.

In a broad sense optimizing production time and ensuring product quality, affected by drying and shrinkage, are the two main challenges of biomaterials such as FLAM. To address those requires research on predictive modeling to estimate mitigations that can be applied proactively [25]. If it is possible to predict the shape a freshly printed object will eventually take, then it may be possible to adjust its design's geometry to incorporate corrections that will ensure the results meet the desired targets (→ Fig. 3b). This may be approached via simulation or experimental modeling methods. Initial attempts included modeling the interactions between the material, namely its mechanical properties and shrinkage characteristics, within its environment, namely the loss of water mass via convective drying. The simulation process was based on finite elements analysis methodology. However, sufficiently characterizing the material properties over time was challenging and the complexity of the model escalated quickly without achieving adequate results. An alternative approach evaluated was based on machine learning methodology where the aim was to first train a model to predict geometric deformations due to shrinkage and then adjust the 3D prints at the machine path level to perform proactive corrections [26]. Surprisingly, despite this approach having no understanding of the complex physical phenomena taking place during 3D printing and material drying, it produced satisfactory results while requiring little computational resources (Fig. 3c). Perhaps using similar predictive-corrective techniques can be employed in the future not only before but also during 3D printing (see Fig. 4).

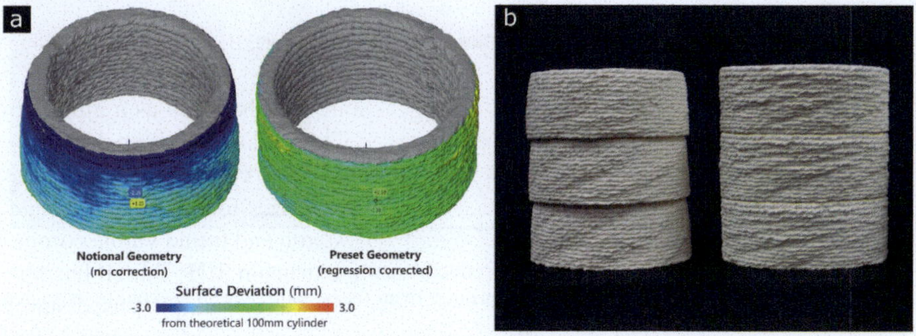

Fig. 4 **a** Overlay of 3D scanned 20-layer cylinders with 100 mm diameter with radial errors and distributions there-of. Shrinkage causes significant surface deflection while modifying the machine path using the machine learning method minimizes errors. **b** Contrasting the effects of the predictive-corrective model using the same three uncorrected and corrected cylinders stacked on top of one another

5 Circular Manufacturing

Despite the abundance and availability of cellulose and chitin around the world, the production of the latter is often concentrated near coastal areas due to its derivation from mainly the fishing industry [27]. On the other hand, cellulose is available even in urban environments in the form of waste streams such as paper and textiles. Environmental analysis of the FLAM products using the Life Cycle Analysis methodology [20], highlighted this discrepancy in the form of transport energy requirements for sourcing chitin. To assess whether FLAM can be produced and consumed locally, without need for transcontinental shipping and transport, we performed material synthesis exclusively from waste streams, and in particular urban waste without dependency on forestry or aquaculture [28]. Food waste is a major problem for municipalities around the world [29] because unlike other waste streams it is challenging to manage contamination or benefit from energy recovery via combustion. Bioconversion is therefore becoming a promising approach for managing urban waste via insect farming [30]. Insects such as the black soldier flies can metabolize multiple times their own weight in food waste [31], and they are used for protein extraction and chemical components including chitin. Demonstration of deriving FLAM from cellulosic and food waste, with equivalent properties to industrial components, thus offers a paradigm for circular manufacturing [32] of bioplastics that can be produced and consumed regionally (Fig. 5).

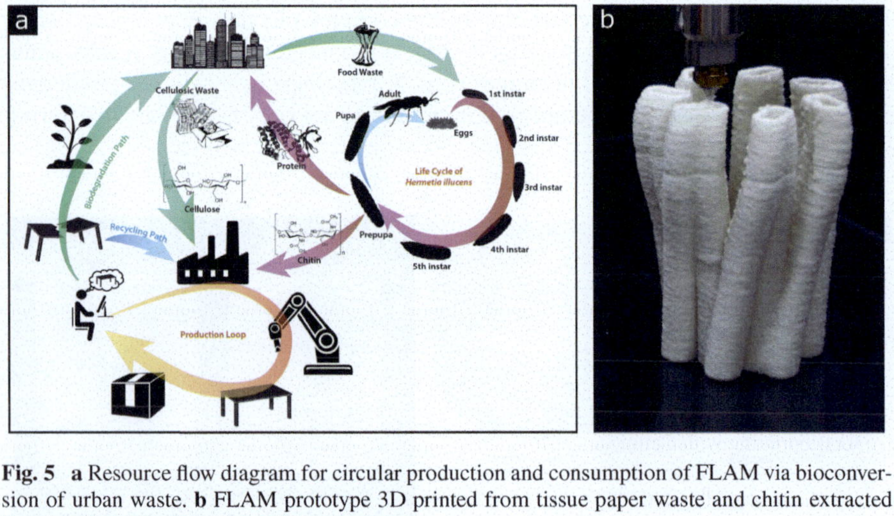

Fig. 5 **a** Resource flow diagram for circular production and consumption of FLAM via bioconversion of urban waste. **b** FLAM prototype 3D printed from tissue paper waste and chitin extracted from black soldier flies fed with food waste

6 Prototypes

The first large-scale FLAM prototype developed was a 1.2 m long airfoil design (Fig. 6a). It was 3D printed in two halves with 50% infill requiring 1 h per piece. The two pieces were fused and coated with FLAM and hand polished. The objective of this prototype was to demonstrate the versatility of the material by showing that it can be used for 3D printing. It has self-adhesive properties evidenced by bonding two cured parts, and that it is compatible with conventional casting and wood-working techniques such as coating and sanding. While the material's surface characteristics were not suitable for airfoil applications, it may be an alternative to current skin-and-core designs which use either natural materials such as balsa wood or synthetic foams such as polyurethanes. To demonstrate the free-form fabrication capability, material strength, assess the reliability of the extrusion system and understand the workflow, we developed an architectural-scale prototype for a pillar with diameter ranging between 0.6 m and 1.0 m and height of 5 m (Fig. 6b). The form was split into 50 vertical segments of 150 mm height each requiring between 30–120 min of 3D printing per segment. The segments were comprised of two adjacent filaments beads with wall thickness of 25 mm. Alternative wall designs were tested such as incorporating buttress fins or an internal web pattern for increased stiffness, but the double-wall design was used as it was the more time efficient configuration for its strength. The amount of time required for printing was 60 h with total wet material weight of 480 Kg while the weight of the pillar after drying was 105Kg. The cost of materials was approximately 360SGD (280 USD) or 0.75 SGD/Kg (0.58 UDS/Kg) in wet and 3.4 SGD/Kg (2.7 USD/Kg) in dry FLAM. Apart from human-related operating errors that required recycling the material and reprinting two segments, the process was robust. The predictive-corrective models used to adjust the design geometry to account for shrinkage gave overall good results with approximately 5 mm or less than 1% error over the diameter of the segments.

7 Potential Applications

FLAM represents a class of materials distinguished by unique properties that are not yet widely understood by designers, engineers, manufacturers, or end-users. Exploring potential applications and stimulating market demand are essential, not only where FLAM serves as a more sustainable alternative but also in enabling innovative product designs. Potential applications include disposable products, furniture and fittings, industrial components, building construction, and cultural uses.

Disposability, while contentious, is nonetheless inherent to various product categories; replacing materials that are difficult to recycle with biodegradable composites such as FLAM holds significant promise. Unlike bioplastics that require industrial composting, FLAM biodegrades naturally and thus suits use-cases where product lifespan is short. Although mass-produced disposable items like packaging typically

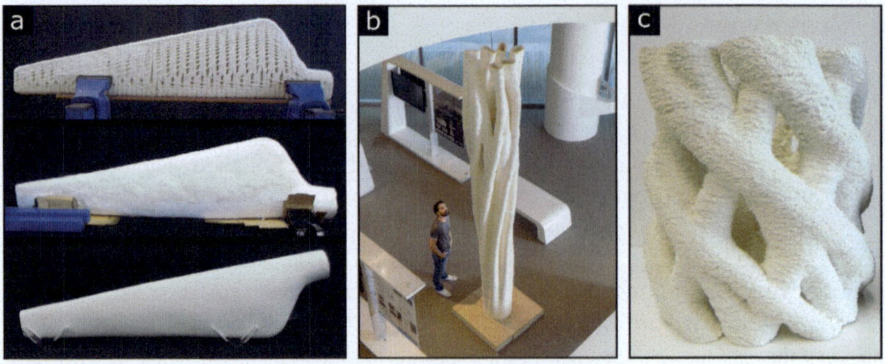

Fig. 6 **a** 1.2 m long airfoil prototype demonstrates the versatility of FLAM which can be 3D printed, used as adhesive or coating and can be hand-finished with wood-working techniques. **b** Architectural prototype demonstrating the ability to build large-scale self-supporting structures (Photography by Frank Pinckers). **c** Rapid prototype demonstrating higher resolution 3D printing without support material and with large overhangs

demand high production rates incompatible with additive manufacturing, 3D printing FLAM is relevant for rapid prototyping, where objects are typically short-lived [33]. The use of FLAM could replace thermoplastics commonly used yet seldom recycled [34] in fused filament fabrication. The current system has demonstrated the ability to print high-resolution, complex shapes with considerable unsupported tapers and layer heights of about 500 μm, a limitation arising from the mechanical system's positional accuracy which may be improved in future iterations.

FLAM's similarity to natural timber enables its application in interior design elements such as fittings and furnishings, addressing the limitations of engineered composite wood plastics commonly used today. Conventional products, including wood laminates and particle or fiber boards, rely on binders such as urea-formaldehyde and phenolic-formaldehyde, which are produced through energy-intensive processes, emit hazardous volatile organic compounds, and transform otherwise natural wood into non-recyclable waste, with adverse health and environmental consequences at end-of-life [35–37]. FLAM, utilizing chitin as a binder, eliminates these drawbacks while meeting mechanical requirements. Additive manufacturing offers value through customization, ranging from minor dimensional adjustments for existing interiors to the creation of entirely bespoke designs with complex geometries.

Several potential applications exist for producing industrial components using FLAM. For example, cellulosic materials are currently used in electrical insulators infused with dielectrics [38, 39], thermal insulation in the automotive and building sectors [40, 41], and lightweight structural sandwich composite panels for aerospace and naval uses [42–44]. The first large-scale object produced with FLAM was a wind blade prototype. This project aimed to demonstrate the feasibility of 3D printing complex shapes for laminate composite core structures, which are traditionally manufactured from balsa wood or synthetic foams such as polystyrene or polyurethane. FLAM offers a lightweight option for these core-and-skin applications, and additive manufacturing allows engineers to adjust the density of the inner structure according to performance requirements.

Additive manufacturing for buildings has been a subject of extensive research [45]. Efforts have focused on pre-fabricating structural parts and directly 3D printing buildings on site using concrete [46]. Several challenges exist regarding the structural properties of 3D printed concrete, mainly because construction grade concrete is not currently suitable for 3D printing [47]. Additionally, reinforcing 3D printed concrete structures presents further difficulties [48], which impacts compatibility with building regulations. Creating forms with FLAM allows for the use of conventional concrete mixtures and enables the embedding of code-compliant rebar cages, which resolves some reinforcement challenges. Further, 3D printing FLAM offers potential advantages in design integration, as the forms can be retained and serve as an insulation layer, potentially reducing the need for additional materials such as polystyrene or polyurethane. FLAM forms are also easier to finish and facilitate the embedding of services such as electrical and mechanical conduits compared to raw concrete.

Since its introduction, FLAM additive manufacturing has attracted interest for cultural uses like heritage conservation, enabling the recreation of historic timber components with 3D printing. As traditional crafts decline and buildings are increasingly digitized, FLAM offers a way to preserve authenticity by using original timber species in restorations. In entertainment, FLAM are a sustainable alternative to foams and plastics for creating lightweight, sculptural props and stage designs, aligning with efforts to reduce polymer use.

8 Conclusions

Design considerations, operational characteristics, and potential applications for a new additive manufacturing technology using cellulosic-chitinous composites aimed at sustainable manufacturing are discussed. While natural biological materials cannot be used universally, they may reduce reliance on synthetic polymers. A significant consideration when introducing a new material process is addressing established approaches shaped by experience with existing technologies. Natural biomaterials exhibit different properties from conventional plastics, and attempts to make them mimic plastics, such as thermoforming or resisting oxygen and water, can affect their ecological features. These biomaterials possess distinct properties that may offer alternative uses. Increasing access to the FLAM additive manufacturing technology for designers, engineers, and manufacturers may facilitate knowledge development and inform product design. Future work includes improving FLAM's material properties for broader application, enhancing process and product quality control to accelerate production, and researching new biomaterials for improved weather resistance and biodegradation.

Acknowledgments The research work was supported by the IDC-IDG21600101, NAMIC-2016026, DManD-RGDM1620303, MOE2018-T2-2-176, and CRP20-2017-0004 grants.

References

1. Reiterer, A., Lichtenegger, H., Tschegg, S., Fratzl, P.: Experimental evidence for a mechanical function of the cellulose microfibril angle in wood cell walls. Philos. Mag. A. **79**(9), 2173–2184 (1999). https://doi.org/10.1080/01418619908210415
2. Klemm, D., Heublein, B., Fink, H., Bohn, A.: Cellulose: fascinating biopolymer and sustainable raw material. Angew. Chem. Int. Ed. **44**, 3358–3393 (2005). https://doi.org/10.1002/anie.200460587
3. Dufresne, A.: Nanocellulose: From Nature to High Performance Tailored Materials. Walter de Gruyter GmbH & Co KG, France (2017). https://doi.org/10.1515/9783110480412
4. Cohen, E.: Chitin biochemistry: synthesis and inhibition. Annu. Rev. Entomol. **32**(1), 71–93 (1987). https://doi.org/10.1146/annurev.en.32.010187.000443
5. Rinaudo, M.: Chitin and chitosan: properties and applications. Prog. Polym. Sci. **31**(7), 603–632 (2006). https://doi.org/10.1016/j.progpolymsci.2006.06.001
6. Morin-Crini, N., Lichtfouse, E., Torri, G., Crini, G.: Applications of chitosan in food, pharmaceuticals, medicine, cosmetics, agriculture, textiles, pulp and paper, biotechnology, and environmental chemistry. Environ. Chem. Lett. **17**(4), 1667–1692 (2019). https://doi.org/10.1007/s10311-019-00904-x
7. French, A.D., Pérez, S., Bulone, V., Rosenau, T., Gray, D.: Cellulose. In: Mark, H.F. (ed.) Encyclopedia of Polymer Science and Technology, pp. 1–69. John Wiley & Sons, Ltd. (2018). https://doi.org/10.1002/0471440264.pst042.pub2
8. Geyer, R., Jambeck, J.R., Law, K.L.: Production, use, and fate of all plastics ever made. Sci. Adv. **3**(7) (2017). https://doi.org/10.1126/sciadv.1700782
9. Liebert, T.F., Heinze, T.J., Edgar, K.J.: Cellulose solvents: for analysis, shaping and chemical modification. Am. Chem. Soc. **1033** (2010). https://doi.org/10.1021/bk-2010-1033
10. Tokiwa, Y., Calabia, B.P.: Biodegradability and biodegradation of poly(lactide). Appl. Microbiol. Biotechnol. **72**, 244–251 (2006). https://doi.org/10.1007/s00253-006-0488-1
11. Atiwesh, G., Mikhael, A., Parrish, C.C., Banoub, J., Le, T.A.T.: Environmental impact of bioplastic use: a review. Heliyon. **7**(9) (2021). https://doi.org/10.1016/j.heliyon.2021.e07918
12. Sanandiya, N.D., Vijay, Y., Dimopoulou, M., Dritsas, S., Fernandez, J.G.: Large-scale additive manufacturing with bioinspired cellulosic materials. Sci. Rep. **8** (2018). https://doi.org/10.1038/s41598-018-26985-2
13. Bartnicki-Garcia, S.: Cell wall chemistry, morphogenesis, and taxonomy of fungi. Ann. Rev. Microbiol. **22**, 87–108 (1968). https://doi.org/10.1146/annurev.mi.22.100168.000511
14. Lévesque, C.A.: Fifty years of oomycetes—from consolidation to evolutionary and genomic exploration. Fungal Divers. **50**, 35–46 (2011). https://doi.org/10.1007/s13225-011-0128-7
15. Fernandez, J.G., Ingber, D.E.: Unexpected strength and toughness in chitosan-fibroin laminates inspired by insect cuticle. Adv. Mater. **24**, 480–484 (2011). https://doi.org/10.1002/adma.201104051
16. Ng, S., Dritsas, S., Fernandez, J.G.: Martian biolith: a bioinspired regolith composite for closed-loop extraterrestrial manufacturing. PLoS One. **15** (2020). https://doi.org/10.1371/journal.pone.0238606
17. Sherin, P., Lyczko, N., Deepu, G., Hanna, M.J., Ange, N., Sabu, T.: Chitin and chitosan based composites for energy and environmental applications: a review. Waste Biomass Valor. **12**, 4777–4804 (2020). https://doi.org/10.1007/s12649-020-01244-6
18. Ng, S., Ng, G.-Z.B., Simpson, R.E., Fernandez, J.G.: a biological approach to metalworking based on chitinous colloids and composites. Adv. Funct. Mater. **34**(51) (2024). https://doi.org/10.1002/adfm.202406800

19. Fratzl, P., Weinkamer, R.: Nature's hierarchical materials. Prog. Mater. Sci. **52**, 1263–1334 (2007). https://doi.org/10.1016/j.pmatsci.2007.06.001
20. Ng, S., Song, B., Fernandez, J.G.: Environmental attributes of fungal-like adhesive materials and future directions for bioinspired manufacturing. J. Clean. Prod. **282** (2021). https://doi.org/10.1016/j.jclepro.2020.125335
21. Hölzl, K., Lin, S., Tytgat, L., Van Vlierberghe, S., Gum, L., Ovsianikov, A.: Bioink properties before, during and after 3D bioprinting. Biofabrication. **8**(3) (2016). https://doi.org/10.1088/1758-5090/8/3/032002
22. Thibaut, C., Denneulin, A., Rolland du Roscoat, S., Beneventi, D., Orgéas, L., Chaussy, D.: A fibrous cellulose paste formulation to manufacture structural parts using 3D printing by extrusion. Carbohydr. Polym. **212**, 119–128 (2019). https://doi.org/10.1016/j.carbpol.2019.01.076
23. Vijay, Y., Sanandiya, N.D., Dritsas, S., Fernandez, J.G.: Control of process settings for large-scale additive manufacturing with sustainable natural composites. J. Mech. Des. **141**(8) (2019). https://doi.org/10.1115/1.4042624
24. Dritsas, S., Hoo, J.-L., Fernandez, J.G.: Sustainable rapid prototyping with fungus-like adhesive materials. CAADRIA. **2**(263–272), 10.52842/conf.caadria.2022.2.263 (2022)
25. Moroni, G., Syam, W.P., Petrò, S.: Towards early estimation of part accuracy in additive manufacturing. Procedia CIRP. **21**, 300–305 (2014). https://doi.org/10.1016/j.procir.2014.03.194
26. Dritsas, S., Ravindran, R., Hoo, J.-L., Fernandez, J.G.: Shrinkage prediction and correction in material extrusion of cellulose-chitin biopolymers using neural network regression. Virtual Phys. Prototyp. **18** (2023). https://doi.org/10.1080/17452759.2023.2225039
27. Younes, I., Rinaudo, M.: Chitin and chitosan preparation from marine sources. Structure, properties and applications. Mar. Drugs. **13**(2), 1133–1174 (2015). https://doi.org/10.3390/md13031133
28. Sanandiya, N.D., Ottenheim, C., Phua, J.W., Caligiani, A., Dritsas, S., Fernandez, J.G.: Circular manufacturing of chitinous bio-composites via bioconversion of urban refuse. Sci. Rep. **10** (2020). https://doi.org/10.1038/s41598-020-61664-1
29. Despoudi, S., Bucatariu, C., Otles, S., Kartal, C.: Food waste management, valorization, and sustainability in the food industry. In: Galanakis, C.M. (ed.) Food Waste Recovery, pp. 3–19 (2021). https://doi.org/10.1016/B978-0-12-820563-1.00008-1
30. Spranghers, T., Ottoboni, M., Klootwijk, C., Ovyn, A., Deboosere, S., De Meulenaer, B., Michiels, J., Eeckhout, M., De Clercq, P., De Smet, S.: Nutritional composition of black soldier fly (Hermetia illucens) prepupae reared on different organic waste substrates. J. Sci. Food Agric. **97**, 2594–2600 (2016). https://doi.org/10.1002/jsfa.8081
31. Gao, Z., Wang, W., Lu, X., Zhu, F., Liu, W., Wang, X., Lei, C.: Bioconversion performance and life table of black soldier fly (*Hermetia illucens*) on fermented maize straw. J. Clean. Prod. **230**, 974–980 (2019). https://doi.org/10.1016/j.jclepro.2019.05.074
32. Fernandez, J.G., Dritsas, S.: The biomaterial age: the transition toward a more sustainable society will be determined by advances in controlling biological processes. Matter. **2**, 1352–1355 (2020). https://doi.org/10.1016/j.matt.2020.04.009
33. Drizo, A., Pegna, J.: Environmental impacts of rapid prototyping: an overview of research to date. Rapid Prototyp. J. **12**, 64–71 (2006). https://doi.org/10.1108/13552540610652393
34. Baechler, C., DeVuono, M., Pearce, J.M.: Distributed recycling of waste polymer into RepRap feedstock. Rapid Prototyp. J. **19**, 118–125 (2013). https://doi.org/10.1108/13552541311302978
35. Piekarski, C.M., de Francisco, A.C., da Luz, L.M., Kovaleski, J.L., Silva, D.A.L.: Life cycle assessment of medium-density fiberboard (MDF) manufacturing process in Brazil. Sci. Total Environ. **575**, 103–111 (2017). https://doi.org/10.1016/j.scitotenv.2016.10.007
36. Jia, L., Chu, J., Ma, L., Qi, X., Kumar, A.: Life cycle assessment of plywood manufacturing process in China. Int. J. Environ. Res. Public Health. **16**(11) (2019). https://doi.org/10.3390/ijerph16112037

37. Nuryawan, A., Risnasari, I., Sucipto, T., Heri Iswanto, A., Rosmala Dewi, R.: Urea-formaldehyde resins: production, application, and testing. In: IOP Conference Series: Materials Science and Engineering, vol. 233 (2017). https://doi.org/10.1088/1757-899X/223/1/012053
38. Prevost, T.A., Oommen, T.V.: Cellulose insulation in oil-filled power transformers: part I—history and development. IEEE Electr. Insul. Mag. **22**, 28–35 (2006). https://doi.org/10.1109/MEI.2006.1618969
39. Hollertz, R., Wagberga, L., Pitois, C.: Novel cellulose nanomaterials. In: Paper Presented at the IEEE 18th International Conference on Dielectric Liquids (2014). https://doi.org/10.1109/ICDL.2014.6893152
40. Loureiro, N.C., Esteves, J.L.: Green composites in automotive interior parts. In: Koronis, G., Silva, A. (eds.) Green Composites for Automotive Applications, pp. 81–97. Woodhead Publishing (2019). https://doi.org/10.1016/B978-0-08-102177-4.00004-5
41. Lopez Hurtado, P., Rouilly, A., Vandenbossche, V., Raynaud, C.: A review on the properties of cellulose fibre insulation. Build. Environ. **96**, 170–177 (2016). https://doi.org/10.1016/j.buildenv.2015.09.031
42. Birman, V., Kardomateas, G.A.: Review of current trends in research and applications of sandwich structures. Compos. Part B Eng. **142**, 221–240 (2018). https://doi.org/10.1016/j.compositesb.2018.01.027
43. Asim, M., Saba, N., Jawaid, M., Nasir, M.: Potential of natural fiber/biomass filler-reinforced polymer composites in aerospace applications. In: Jawaid, M., Thariq, M. (eds.) Sustainable Composites for Aerospace Applications, pp. 253–268 (2018). https://doi.org/10.1016/B978-0-08-102131-6.00012-8
44. Pandey, J.K., Ahn, S.H., Lee, C.S., Mohanty, A.K., Misra, M.: Recent advances in the application of natural fiber based composites. Macromol. Mater. Eng. **295**, 975–989 (2010). https://doi.org/10.1002/mame.201000095
45. Khoshnevis, B.: Automated construction by contour crafting—related robotics and information technologies. Autom. Constr. **13**, 5–19 (2004). https://doi.org/10.1016/j.autcon.2003.08.012
46. Paolini, A., Kollmannsberger, S., Rank, E.: Additive manufacturing in construction: a review on processes, applications, and digital planning methods. Addit. Manuf. **30** (2019). https://doi.org/10.1016/j.addma.2019.100894
47. Bos, F., Wolfs, R., Ahmed, Z., Salet, T.: Additive manufacturing of concrete in construction: potentials and challenges of 3D concrete printing. Virtual Phys. Prototyp. **11**, 209–225 (2016). https://doi.org/10.1080/17452759.2016.1209867
48. Panda, B., Tay, Y.W.D., Paul, S.C., Tan, M.J.: Current challenges and future potential of 3D concrete printing: Aktuelle Herausforderungen und Zukunftspotenziale des 3D-Druckens bei Beton. Mater. Werkst. **49**, 666–673 (2018). https://doi.org/10.1002/mawe.201700279

Stylianos Dritsas is an Associate Professor and Associate Head of Pillar (Education) in Architecture and Sustainable Design (ASD) at the Singapore University of Technology and Design (SUTD). His research interests are in design computation, digital fabrication and sustainable manufacturing. He developed the Computer Aided Design and Manufacturing software and the robotic material extrusion system for the large-scale FLAM bio-composites 3D printing system. In the past he taught at the Harvard Graduate School of Design, Ecole Polytechnique Federale in Lausanne, the Architectural Association in London, and practiced architecture at dECOi Architects in Massachusetts and Kohn Pedersen Fox Associates in London.

Javier G. Fernandez is an Associate Professor in Engineering Product Development (EPD) at the Singapore University of Technology and Design (SUTD). His research is on the study of biological materials. At SUTD, he designed FLAM, a cellulosic bio-composite for large-scale manufacturing; at the Harvard Wyss Institute he designed Shrilk, a biocompatible material inspired by the insect cuticle; and at MIT, he developed MicroMasonry, a technology to assemble artificial organs.

He was awarded the world's most outstanding young researcher in materials science by the Bayer Foundation, the Zwick Science Award for his studies on Mechanical Testing, and best PhD thesis at the University of Barcelona.

Chapter 5
3D Bioprinting with Cellulose-Derived Bio-ink and Precursors

Bashir Khoda and Md. Ahasan Habib

Abstract Three-dimensional (3D) bioprinting has revolutionized tissue engineering by enabling the fabrication of functional living structures through the precise layer-by-layer deposition of cell-laden biomaterials. However, challenges persist in achieving mechanical integrity, maintaining structural fidelity, and ensuring desired cellular activities, including viability, proliferation, and differentiation, in printed constructs. This chapter focuses on three critical aspects of extrusion-based bioprinting: (i) biomimetic design, (ii) bio-inks and precursors characterization, and (iii) printing with cellulose and its derivatives. Human organs consist of heterogeneous cell populations organized into gradient architectures within a variably porous extracellular matrix (ECM). Replicating such intricate structural patterns necessitates customized design and deposition strategies tailored specifically to cells, tissues, and organs—topics explored in the first section. Cellulose-based biomaterials, particularly carboxymethyl cellulose (CMC) and TEMPO-mediated nanofibrillated cellulose (TO-NFC), have become promising candidates for bio-inks due to their intrinsic biocompatibility, minimal cytotoxicity, and adjustable rheological characteristics. The second section details methodologies for preparing cellulose-derived bio-inks and precursor formulations. Finally, the third section analyzes the influence of printing parameters on structural fidelity and functional performance, emphasizing their role in cellular encapsulation efficacy and sustained viability within biofabricated living constructs. By integrating the troika (design, material, and manufacturing) factors in our bioprinting methodology, we demonstrate that cellulose-based bio-inks can effectively produce large-scale structures with high geometric fidelity at low solid concentrations (<5%). Furthermore, cell encapsulation using these bio-inks achieves sustained viability, with reported cell survival rates ranging from 86–90% over extended incubation periods. Our study serves as an essential reference for researchers and practitioners seeking to leverage cellulose derivatives in next-generation, biodegradable 3D bioprinting applications.

B. Khoda (✉)
Mechanical Engineering Department, University of Maine, Orono Campus, Orono, ME, USA
e-mail: bashir.khoda@maine.edu

Md. A. Habib
Manufacturing and Mechanical Engineering Technology, Rochester Institute of Technology, Rochester, NY, USA

© The Author(s), under exclusive license to Springer Nature Singapore Pte Ltd. 2026

C. Klemmt and G. Beaucage (eds.), *Biodegradable 3D-Printing*, The Materials Research Society Series, https://doi.org/10.1007/978-981-95-4025-9_5

Keywords Bioprinting · Porosity · Biomimetic · Biotopology · Hydrogel · Bio-ink

1 Introduction

3D bioprinting, a specialized domain within additive manufacturing (AM), leverages the layer-by-layer fabrication approach to construct complex biological structures [1, 2]. This technology enables the creation of highly porous 3D scaffolds that serve as temporary structural supports for growing cells, providing nutrients to newly forming tissues, facilitating the healing process, restoring tissue function, and minimizing wound scarring [3–6]. Traditionally, tissue engineering relies on preformed biodegradable polymeric scaffolds fabricated using methods such as casting, particulate leaching, gas foaming, phase separation, melt molding, freeze-drying [7, 8], and electrospinning [9]. These techniques are process-driven rather than design-driven and are often limited in achieving complex geometry and internal architecture. These limitations can lead to challenges in fabricating uniform or functionally graded heterogeneous cellular structures, reducing repeatability and causing non-uniform fluid flow, uncontrolled permeability, or unplanned degradation, which can hinder tissue regeneration. In contrast, 3D bioprinting provides precise spatial control and repeatability of material deposition, enabling the fabrication of design-specific tissue scaffolds. This makes 3D bioprinting an ideal platform for achieving biologically inspired, physiologically relevant, and computer-assisted scaffold designs that promote tissue regeneration (Fig. 1) [10, 11].

Like traditional AM processes, 3D bioprinting begins with a digitized model of an organ or tissue which is often obtained through reverse engineering techniques, e.g., computed tomography (CT) scans or magnetic resonance imaging (MRI) to achieve the physiologically accurate geometry. In traditional AM, only the surface information of the models are further processed and saved as Stereolithography (STL) files, while material and topological representations are often ignored or deemed redundant. These monochrome STL files represent the model's outer surface as a

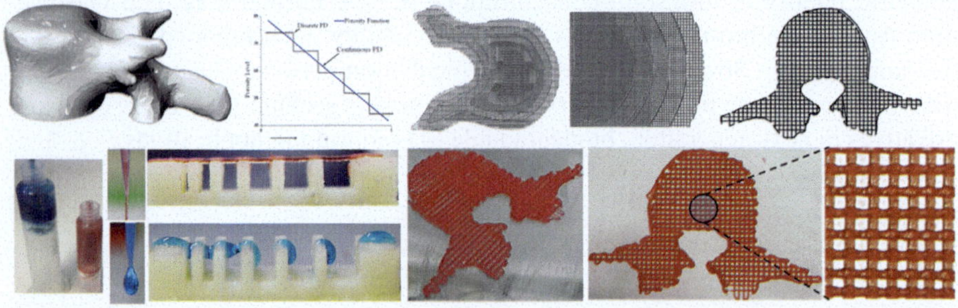

Fig. 1 Graphical abstract

tessellated mesh of triangular approximations [12]. During preprocessing, the STL model is sliced into thin layers by intersecting planes perpendicular to a predefined build direction [13]. The slicing process generates contours for each layer, forming the foundation for toolpath generation, which dictates material deposition and support structure creation. By stacking these layers sequentially along the build direction, a three-dimensional physical model is realized.

3D bioprinting follows a similar workflow but incorporates bio-inks—combinations of cells, biomaterials, and growth factors—to create tissue-like porous structures often known as scaffold structures [14, 15]. For successful tissue regeneration, scaffolds must meet several criteria, including supporting cell migration, proliferation, and differentiation, enabling the removal of waste products, and promoting vascularization [16, 17]. Thus, material and topological information are often considered significant to achieve biomimicry and functional living structures for a successful bioprinting process. Data digitization, materials selection and manufacturing/fabrication are the three major steps in the bioprinting process (cf. Fig. 2). The data digitization can also be referred to as preprocessing and primarily involves Computer Aided Design-based techniques to achieve physiological boundaries and internal features. Simultaneously, a suitable biocompatible material is designed with shear-thinning behavior and sufficient polymerization for self-supporting characteristics while maintaining enough pores for cells to interact and grow. By incorporating and utilizing this information, a robust manufacturing process plan with defined machine motion for bio-ink deposition patterns is developed which plays a crucial role in determining the printability, shape fidelity, and functionality of the fabricated constructs [18–20].

This chapter will focus on the troika (design, material, and printing) effect on 3D bioprinting, as those are highly interconnected in guiding the entire biofabrication workflow. The topics are designed and discussed related to extrusion-based bioprinting with cellulose materials.

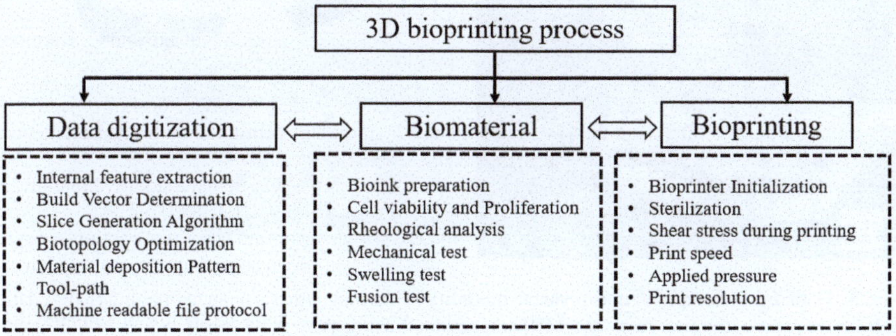

Fig. 2 Systematic approach to 3D bioprinting processes

2 Data Digitization

2.1 Reverse Engineering

Scaffold architecture, particularly porosity, plays a crucial role in cell survivability, proliferation, and migration [21, 22]. Porous scaffolds have been shown to improve cell viability by almost 40% compared to non-porous scaffolds. Although traditional CAD systems are the most common tools for digitizing models for 3D bioprinting, successful tissue engineering applications require internal spatial density and porosity information that can be achieved through reverse engineering (RE) techniques. RE is the process of duplicating an existing part, sub-assembly, or product, without drawings, documentation, or a computer model, which is the appropriate tool for free-form geometry such as organs or body parts. In this chapter, we will discuss the reconstruction of geometry and hidden details with non-invasive and non-contact modalities such as CT, micro-CT, or magnetic resonance imaging (MRI) data. The fundamental process depicted in Fig. 3 involves a series of overlapping phases that often require multiple iterations to refine and achieve the desired digitized model. This iterative approach ensures precision and accuracy in replicating the original geometry, topology, and internal spatial features which are essential for biomimicry.

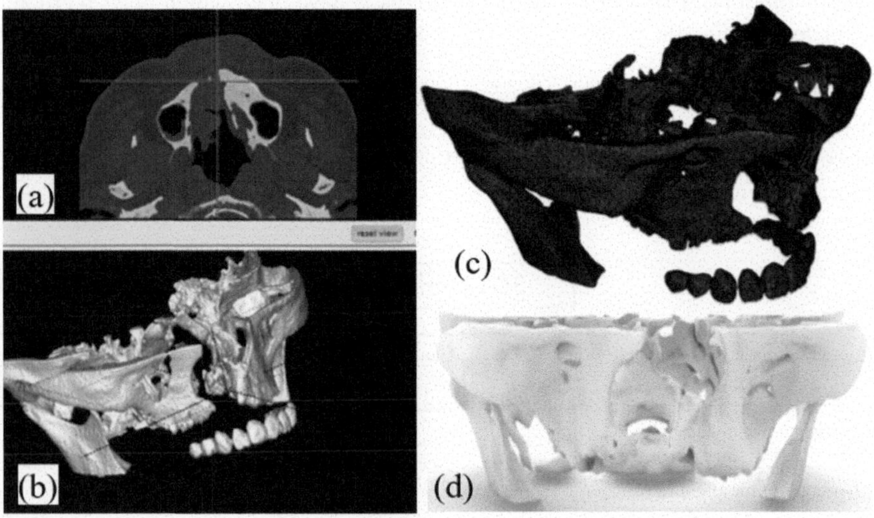

Fig. 3 Data digitization with non-invasive modality using magnetic resonance imaging (MRI) data **a–b** hard tissue extraction from the MRI image stack using gray scale thresholding in ITK snap software; **c** analyzing and fixing the digital model in the CAD platform **d** proper reconstruction of the biomodel with internal features (i.e., density and porosity) information

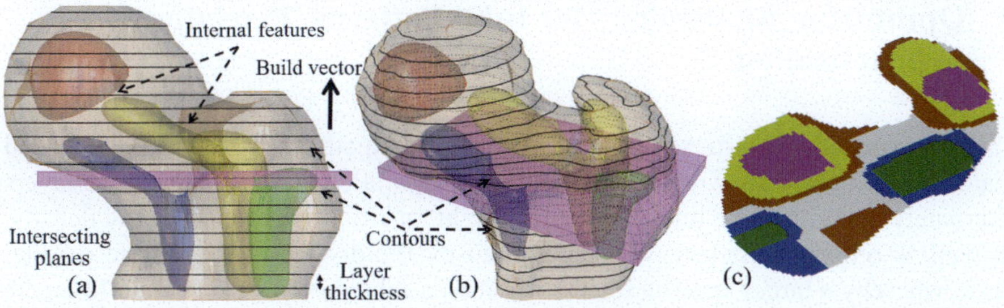

Fig. 4 Image illustrates the slicing process and contour generation with internal features **a–b** biomodel with internal features are sliced along the build vector and **c** generated layer with internal features information as color map

2.2 Slicing Biomodels Along the Desired Build Direction

The biomodel is sliced into layers along the desired build vector for further processing toward bioprinting. The build direction is typically represented by a perpendicular vector on an imaginary plane, guiding the layer-by-layer material deposition [23, 24]. It can also be viewed as the augmentation vector connecting the bottom and top layers during fabrication. This orientation determines the cross-sectional area where material is deposited, directly influencing the structural properties, heterogeneity of the manufactured part and cannot be decoupled from the overall plan for the 3D bioprinting process. A set of parallel slicing planes, perpendicular to the build direction, intersect with the mesh model and mark the boundaries where the planes meet the facets, forming boundary and internal contours in each layer, preserving the feature information of the biomodel (Fig. 4a-b). The distance between parallel slicing planes, known as layer thickness, remains uniform in standard slicing techniques.

2.3 Toolpath

The toolpath, also known as the deposition or scan pattern, represents the final step in the AM preprocessing workflow. For each layer, a toolpath is generated for the contour as boundary and space filling infill deposition plan. However, interconnected porosity is essential in 3D bioprinting due to the ease of transportation of nutrients, biomolecules and cells. Thus, the space filling infill deposition pattern is often a focus with defined deposition angles, and raster width (RW: filament to filament distance) to fill the contours. Commonly, a crisscrossing or zigzag raster pattern of 0°–90° is applied across successive layers, enhancing structural integrity. To address biomimicry in the bioprinting process, internal features (i.e., porosity, density) need to be captured with variational patterned infill and this research field can be defined as biotopology optimization. One such technique named "Pixelation to Voxelization" is discussed here.

3 Optimizing Biotopology for Infill Patterns: Pixelation to Voxelization

Task Definition: The aim is the development of a toolpath for continuously extruded material. The toolpath will follow a pattern of parallel lines, with vertical layers alternately printing those lines from left to right, or from front to back. The density, as well as other parameters, will be defined by the distance between the parallel lines, whereby a larger spacing leads to an increased porosity of the print. To create a functionally graded print with different densities across its volume, the spacing of the lines needs to be optimized to best match the intended distribution of the varying material densities.

The process starts with a sliced layer containing the internal features related to biomimicry with continuum or discretely distributed heterogeneity as shown in Fig. 4c. Such images can also be generated from microscopic images with RGB or gray scale maps related to internal features, or constructed from CAD tools, FEA analysis, topology or topometry optimization tools, or can be the cross-sectional image of the object. Figure 5 shows an example of discrete heterogeneity which can be represented with the following expression (Eq. 1):

$$\begin{aligned} &\mathbf{L} = [\mathbf{F}, \mathbf{C}]\,; \\ &\mathbf{F} = \{\mathbf{F}_f = \{M_m, L_l, A_v,\ P_{f,1}..P_{f,n},\ C_f\}\}_{f=0\ldots F} \\ &\forall f;\ m, l, v \in Z \end{aligned} \tag{1}$$

where $\mathbf{F}$ is the set of features in heterogeneous layer $\mathbf{L}$; Each feature $\mathbf{F}_f$ is an aggregate of homogeneous pixels and defines a physical boundary. For continuum heterogeneity, the physical boundary of an individual pixel defines the feature. Each feature

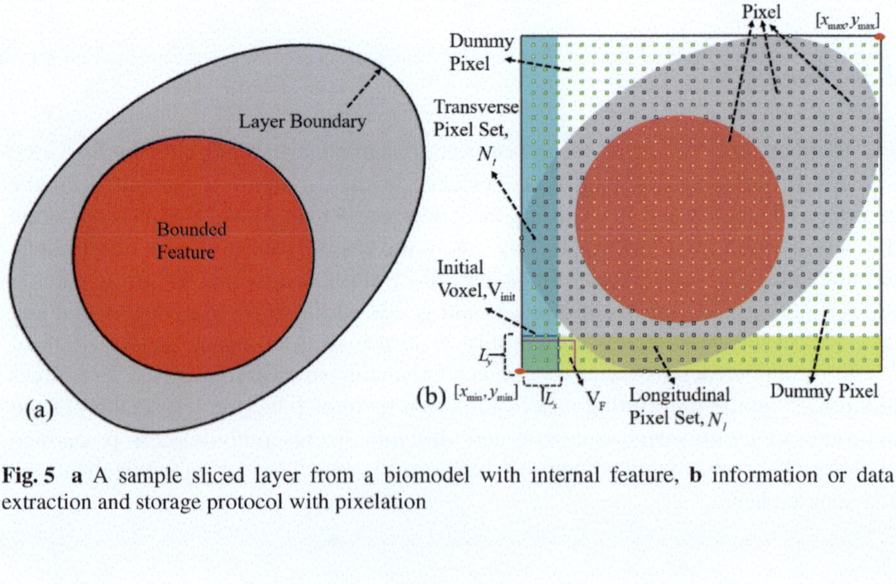

Fig. 5 **a** A sample sliced layer from a biomodel with internal feature, **b** information or data extraction and storage protocol with pixelation

contains characteristics such as material M_m; location L_l; area/volume A_v and other physical characteristics or performance such as density and porosity $P_{f,n}$. C_f represents the constraint sets among the defining characteristics of the feature $\mathbf{F}_f$. Each of these features $\mathbf{F}_f$ can be discretized into small areas defined by pixels **N** (color or gray scale) and share the same characteristics or measured performance (i.e., material distribution/property/type, deformation, strain energy, elastic modulus, etc.) of the parent feature $\mathbf{F}_f$ as shown in Fig. 5.

A bounding box is generated for the domain following the boundary equations: $x = x_{\max}; y = y_{\max}; x = x_{\min}; y = y_{\min}$ where, $\mathrm{X} = [x_{\min}, x_{\max}]$ and $\mathrm{Y} = [y_{\min}, y_{\max}]$ define the range of the domain. Dummy pixels, N_d with null properties are defined for the domain that is not occupied by features $\mathbf{F}_f$ as shown in Fig. 5. An iterative diagonal voxel construction technique will be used where the voxel will initiate from $[x_{\min}, y_{\min}]$ as shown in Fig. 5. The attributes of the voxel will be determined by the neighboring pixels. To match the design domain-dimensions with the manufacturing capabilities, only the longitudinal and transverse pixels are considered as neighboring pixels. Each voxel can be represented with both extrinsic and intrinsic property such as $\mathrm{V_{init}} = [L_x, L_y, E_1.....E_n]$ shown in the flow diagram Fig. 6. Here, L_x and L_y represent the length and width while E_i represents intrinsic properties related to the measurable performance (e.g., density and porosity). L_z will be added for 3D voxel construction. The length of the voxel L_x is a function of the attributes of the pixels along the transverse direction N_t ; whereas pixels along the longitudinal direction N_l determines the voxel width L_y, which is define as a noise function shown in Fig. 6. Similarly, the intrinsic property E_i for the voxel can be determined by the same pixel sets $\{N_t, N_l\}$ and the resultant voxel becomes $\mathrm{V_F}(\widehat{L}_x, \widehat{L}_y, \widehat{E}_1 \ldots \widehat{E}_n) \ni \mathrm{V_{init}} \rightarrow \mathrm{V_F}$.

The intrinsic property $\widehat{E}_i$ of the resultant voxel $\mathrm{V_F}$ is then mapped back into the corresponding property set $\{M_F, L_F, A_F, P_{F,1}..P_{F,n}, C_F\}$ of pixel $\{N_t, N_l\}$. As a result, the neighboring pixel becomes homogenous pixel set $\{N_F\} \ni (N_t, N_l) \rightarrow N_F$ as shown in Fig. 7a. The deviation or error between the transformed neighboring pixel N_F and the actual $\{N_t, N_l\}$ will be measured by Eq. 2 and this will create an

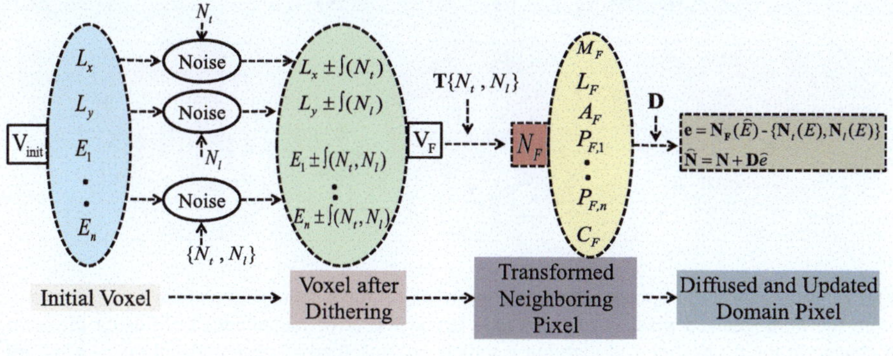

Fig. 6 Process flow map of conformal voxel construction

error map as shown in Fig. 7b (circle diameter depict the error magnitude). The error will be diffused toward the longitudinal and transverse direction through the pixels using a diffusion matrix **D** using Eq. 3. The process flow is simulated in Fig. 7.

$$\mathbf{e} = \mathbf{N_F}(\widehat{E}) - \{\mathbf{N}_t(E), \mathbf{N}_l(E)\} \tag{2}$$

$$\widehat{\mathbf{N}} = \mathbf{N} + \mathbf{D}\widehat{e} \tag{3}$$

Here, **e** is the error measure between the intrinsic property (E) of transformed pixel N_F and the actual neighboring pixel $\{N_t, N_l\}$. **D** is the diffusion matrix that will be used to diffuse the error among non-neighboring pixels **N**; and $\widehat{\mathbf{N}}$ is the updated pixels of the domain at the end of this iteration. In this way, each voxel will contribute and update the accumulating pixels while directly affecting the attributes of the diagonally constructed next voxel as shown in Fig. 7.

Noise Function: Each pixel contains characteristics from its parent feature which can be material (M); location (L); area/volume (A); performance measure (P) (i.e., density, deformation, strain energy, internal energy density, applied load or shape) as well as some other domain-based constraint (C). Each voxel is constructed with the intrinsic and extrinsic properties conforming to the neighboring pixels through a noise function which can be expressed by the following equation (Eq. 4).

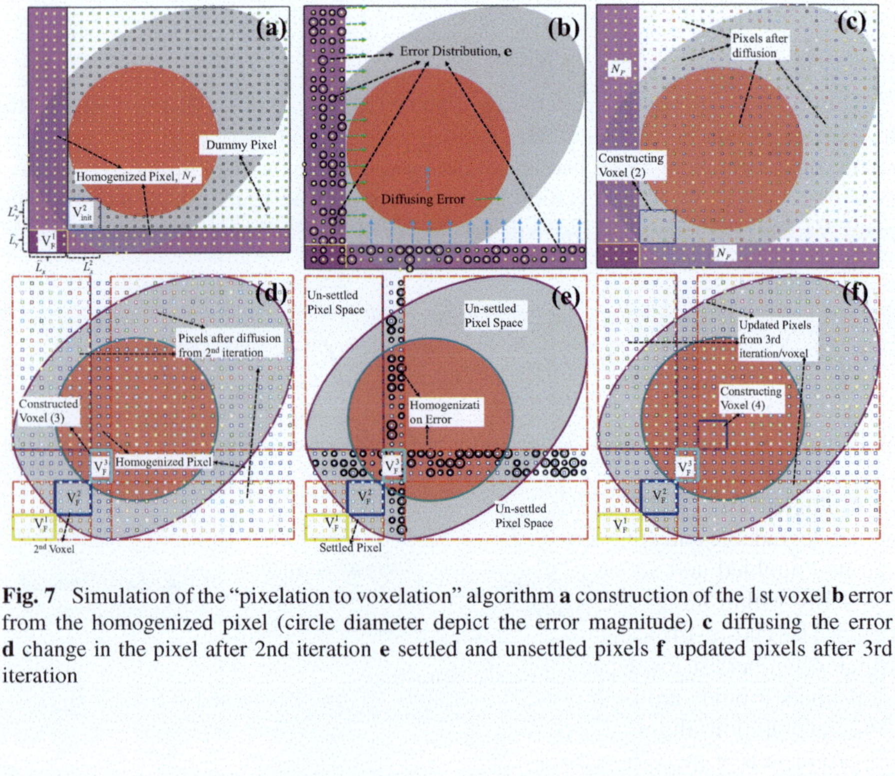

Fig. 7 Simulation of the "pixelation to voxelation" algorithm **a** construction of the 1st voxel **b** error from the homogenized pixel (circle diameter depict the error magnitude) **c** diffusing the error **d** change in the pixel after 2nd iteration **e** settled and unsettled pixels **f** updated pixels after 3rd iteration

$$\mathbf{V_F} = \mathbf{w}_t \mathbf{N_t^T} + \mathbf{w}_l \mathbf{N_l^T} \ni \sum_1^t w_t + \sum_1^l w_l = 1 \tag{4}$$

Here, t and l the numbers of transverse and longitudinal pixel.$\mathbf{V_F}$ is the final voxel with k characteristics represented in a matrix with an order of $k \times t \times 1$, where k is the total number of characteristics considered. $\mathbf{w}_t$ and $\mathbf{w}_l$ are the weights assigned to the corresponding transverse and longitudinal pixel represented in matrix with an order of $1 \times t$ and $1 \times l$ respectively. $\mathbf{N_t}$ and $\mathbf{N_l}$ represent the pixel matrix with an order of $t \times k$ and $l \times k$, respectively.

The appropriate weights assigned to the attributes of the pixels in Eq. 4 can be determined empirically or by using the hypothetical equivalents and inequivalents method (HEIM) [25]. HEIM is mathematically sound technique and can determine the actual importance of the attributes using a set of preferences stated by a decision maker rather than choosing weights arbitrarily from experience or intuition [20]. In this technique, the weight becomes the variable and a set of outcomes is selected from the user preferences (both equality and inequality) as hypothetical alternatives. The user preference between two hypothetical alternatives can be used to form a constraint. Thus, a set of constraints is formulated from the preference information among all the hypothetical alternatives. The objective function in Eq. (5) ensures the summation of all the weight values is unity.

$$\begin{aligned} &\text{Min}\left[1 - \left(\sum_1^t w_t + \sum_1^l w_l\right)\right]^2 \\ &\text{s.t.} \\ &V_{F,i} - V_{F,j} \le -\delta; \quad i, j = 1, 2, \ldots, a \text{ and } i \ne j \\ &0 \le w_t \le 1; \quad \forall t \\ &0 \le w_l \le 1; \quad \forall l \end{aligned} \tag{5}$$

Here, $V_{F,i}$ and $V_{F,j}$ are the values of ith and j^{th} hypothetical alternatives, respectively. δ is a small positive number, which is used to ensure the inequality between two hypothetical alternatives. Solving this equation will yield the characteristics of the voxel as well as homogenize the neighboring pixel.

Diffusion Matrix Construction: The homogenization of the neighboring pixel during voxel construction will introduce error in the domain and can be diffused among the unsettled pixels using a diffusion matrix. The settled pixels in this methodology are pixels that are already assigned to a voxel as shown in Fig. 7e. They will not be considered as neighboring pixels during the current or future voxel construction. Following this methodology, voxels will be constructed throughout the domain covering all pixels including dummy pixels. For shape compatible structures, the dummy pixel will be ignored at the end, but needs to be considered during the noise function and diffusion matrix construction. After the construction of each voxel, the system's error will be determined based on the deviation between settled and actual pixel which can be measured by using the sum square error equation (Eq. 6):

$$SE_i = \sum_x \sum_y (\mathbf{N}_{\mathbf{V}_F^i} - \mathbf{N})^2 \tag{6}$$

Here, SE_i is the system error after i^{th} voxel; $\mathbf{N}_{\mathbf{V}_F}$ represents the pixel matrix in i^{th} voxel $\mathrm{V}_{\mathrm{F}}^i$. By using the *"parallel error diffusion method"* to minimize the system error in each iteration, the unsettled pixel will be readjusted and can provide better feedback for the conforming voxel. In a traditional error diffusion dithering method, the error is diffused to its adjacent or near adjacent neighbor following a serpentine or spiral path weightage, i.e., Floyd and Steinberg [26]. However, Fick's diffusion law can be used to determine the error diffusion weight that can facilitate the manufacturability of the generated pattern. Fick's law establishes the relationship between rate of diffusion and the concentration gradient of pixels, which can be expressed by the following equation:

$$J = -DA(\frac{\Delta e_{N_F^i}}{\Delta X} + \frac{\Delta e_{N_F^i}}{\Delta Y}) \tag{7}$$

Here, J is the diffusion rate or flux; A is the area interface; Δe is the error gradient; X and Y are the directions. The proposed problem can be formulated with Eq. (8) and the voxels are constructed by solving the equation as shown in Fig. 8

$$\begin{aligned} &\text{To Minimize } SE_i - SE_{i-1} \\ &\text{Subject to :} \\ &\frac{\Delta e_{N_F^i}}{\Delta X} \geq 0 \text{ and } \frac{\Delta e_{N_F^i}}{\Delta Y} \geq 0 \end{aligned} \tag{8}$$

Optimizing the Dithering Directions: The characteristics of the generated voxel will depend on their building direction, which is also the dithering direction along the Cartesian coordinate system. By changing the dithering direction, the neighboring pixel set can be changed along all directions. Thus, the conformity of the voxel is a function of the dithering direction and needs to be considered during modeling as shown in Eq. (9).

$$\begin{aligned} &\text{Min} \sum_i SE_i(\alpha, \beta, \gamma) \\ &\text{s.t.,} \frac{\pi}{2} \leq \alpha \leq \frac{3\pi}{2};\ 0 \leq \beta \leq 2\pi; 0 \leq \gamma \leq \pi \end{aligned} \tag{9}$$

Here, α, β, γ defines the three axis of rotation and are the decision variable for the 3D voxelization. For larger solution spaces, stochastic evolutionary search techniques such as genetic algorithms can be used for convergence [20]. The outcome from this biotopology optimization technique can be stored as a mathematical function which

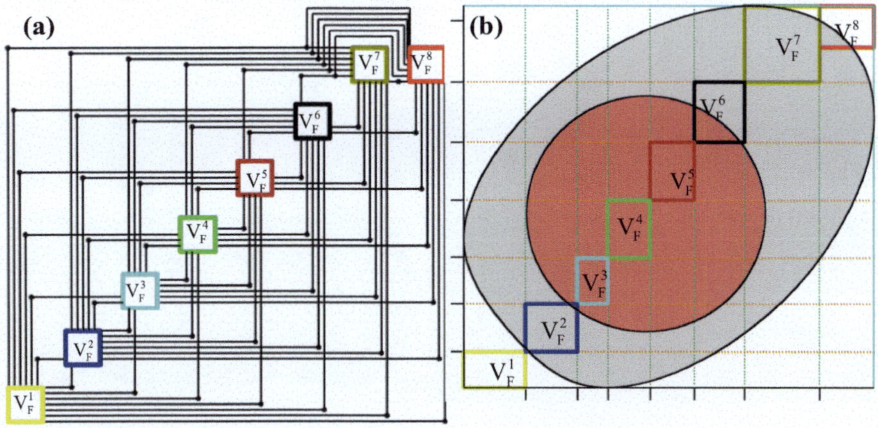

Fig. 8 **a** Interconnected network of voxel construction **b** projected voxel

can be implemented on the layer stack to generate the infill pattern with a continuous deposition plan resembling the internal features of the biomodel and hence achieve biomimicry. A sample layer with three internal features is used to demonstrate the bi-layer variational pattern generation process in Fig. 9. By fitting various unit cells in the voxel, other functional and variational infill patterns with a continuous deposition plan can be generated following this method, which is also shown in Fig. 10.

Machine Readable File for Bioprinters: Once the optimized infill pattern is generated, all essential digital information needs to be fed to the bioprinter for fabrication. G-code stands out as a universally versatile machine-readable file protocol, particularly favored in open-source bioprinters due to its accessibility and customizability. G-code provides precise instructions for controlling the printer's movements, including the deposition path, extrusion rate, and material curing, allowing users to execute specific deposition plans tailored to their needs (e.g., biomimicry). This flexibility is crucial in bioprinting applications, where achieving the desired porosity in a scaffold is critical. While commercially available slicers (i.e., CURA) often generate

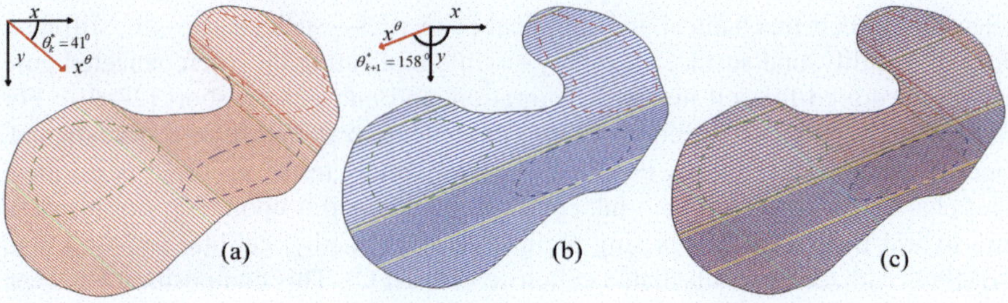

Fig. 9 Bi-layer variational pattern generation process using the "Pixelation to Voxelization" technique. **a–b** Optimized consecutive layer and **c** stacked bi-layer infill pattern

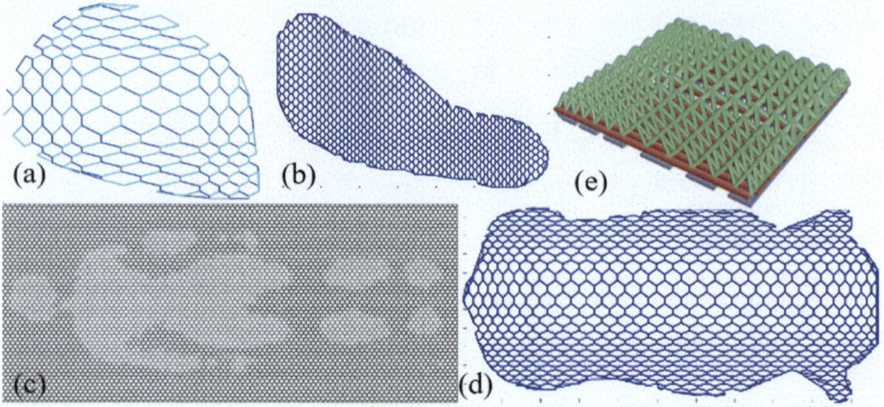

Fig. 10 Variational infill pattern generation process using the "Pixelation to Voxelization" technique. Variational honeycomb pattern for **a** bone scaffold mimicking porosity **b** shoe sole mimicking pressure-point distribution **c** mattress mimicking body mass distribution **d** variation with trilayer and **e** 3D variational lattice structure infill pattern

different infill patterns, the link between an optimized biotopology infill pattern and a slicer generated pattern may not be adequate for meeting the unique mechanical and biological requirements with the scaffold architecture tailored to specific patients or tissues. Customizing the G-code allows researchers and practitioners to adjust toolpaths to fine-tune porosity, density and multiple material amounts ensuring optimal mechanical properties and creating a microenvironment conducive to cellular growth and tissue regeneration. A simple toolpath generation technique using the open-sourced community based "Scratch" platform (https://scratch.mit.edu/) is shown in the supplementary section of this book Appendix.

4 Materials for Bioprinting

To meet the multi-functional requirements, scaffold materials must be carefully selected for their mechanical and chemical properties—such as strength, stiffness, biodegradability, and surface chemistry—while replicating the interconnected pore network required for cell ingrowth, migration, and nutrient transport [27–30]. For an appropriate material selection, a series of characterization tests, e.g., rheological, mechanical, microstructural, chemical, and biological tests, are often performed. This chapter focuses on the bio-ink composition and builds upon our prior research into hybrid hydrogels, comprising alginate, carboxymethyl cellulose (CMC), and TEMPO-oxidized nano-fibrillated cellulose (TO-NFC). This composition has been empirically validated for its biocompatibility—demonstrating approximately 91% cell viability—alongside its structural integrity, capable of supporting a scaffold structure 42 layers high and 1 cm tall, and high printability, evidenced by a printability factor greater than 0.88. These findings have been thoroughly documented in our preceding publications [31].

Alginate: Sodium alginate, a naturally derived polysaccharide extracted from seaweed, has been widely used in 3D bioprinting due to its excellent biocompatibility. Its gelation process is facilitated by the addition of calcium ions, which cross-link the carboxylate groups in sodium alginate, forming a stable gel without compromising the biocompatible environment [32]. The high-water content of alginate hydrogels supports efficient exchange of nutrients and waste products, making it an ideal material for cell encapsulation. Various cell types, including fibroblasts [33], myoblast, endothelial [34], chondrocytes [35], and Schwann [36] have been successfully encapsulated and cultured in alginate-based hydrogels. Under physiological conditions, the gradual dissipation of $Ca2^{+}$ ions render alginate water-soluble and non-toxic, leading to the progressive degradation of calcium alginate gels [37]. However, alginate, despite its widespread use as a bio-ink, has limited effectiveness due to the absence of bio-ligands necessary for mammalian cell attachment [38]. Consequently, achieving an optimal balance of printability, shape fidelity, and biocompatibility using pure alginate hydrogel remains a significant challenge. Balancing these conflicting properties such as mechanical integrity and cellular activities (e.g., viability, proliferation, differentiation, migration, etc.) are crucial for optimizing sodium alginate as a bio-ink in 3D bioprinting applications. Therefore, other hydrogel materials such as gelatin [39], methylcellulose [40], and nanocellulose [41] are often added to achieve better functionality in the fabricated scaffolds.

Carboxymethyl Cellulose (CMC): Carboxymethylcellulose (CMC) is an anionic water-soluble bio-copolymer of β-D-glucose and β-D-glucopyranose-2-O-(carboxymethyl)-mono-sodium salt connected via β-1,4-glucosidic bonds [42]. The presence of carboxyl groups makes this cellulose derivative more soluble, able to thicken, and stable. CMC is a high-molecular-weight, water-soluble polysaccharide widely used as a viscosity modifier or thickener which is also non-toxic and non-allergenic. Research has shown that the matrix-binding properties of CMC support cell migration and attachment [43]. Additionally, hybrid hydrogels composed of alginate and CMC (alg-CMC) have been employed to fabricate beads for various drug delivery applications [44, 45]. The potential of alg-CMC hydrogels for 3D bioprinting is also studied in our previous work demonstrating shape-retention capability, and high cell viability during and after printing [19].

TEMPO-NFC (TO-NFC): Nano-fibrillated cellulose (NFC) is a high-aspect ratio, natural, sustainable material derived from the most abandoned renewable polymer on earth, i.e., plants. NFC is composed of β (1–4) linked D-glucose units and is partially to fully crystalline exhibiting remarkable mechanical properties [46, 47], hydrophilicity and tunable surface chemistries and biocompatibility [41, 48]. Typically, the individual fibrils of NFC are 3–5 nm in diameter and several micrometers in length [49]. Incorporating nano-scale reinforcements such as nano-fibrillated cellulose (NFC) into base hydrogel materials have been explored to enhance both mechanical strength and biological properties, such as cell growth [50, 51]. For instance, human-induced pluripotent stem (iPS) cells have been combined with alginate-NFC and hyaluronic acid-NFC hydrogels to fabricate cartilage tissue constructs [51]. However, pure NFC gel presents challenges during extrusion-based 3D printing due to the non-uniformity of its network entanglements, which causes erratic pressure requirements and limits printability due to clogging [52]. To address these limitations, the surface of NFC is chemically modified through oxidation using 2,2,6,6-tetramethyl-1-piperidinyloxy (TEMPO). This process introduces negatively charged carboxylate groups, creating TEMPO-oxidized NFC (TO-NFC). Chemical treatment significantly enhances the uniformity, dispersibility, and homogeneity of the NFC gel, improving its printability. Furthermore, the structural chains of TO-NFC form mechanically robust crystalline networks through hydrogen bonding, making it a highly suitable bio-ink for fabricating 3D scaffold materials. Finally, adding the high-aspect ratio of the nanofiber into the alginate matrix increases its fibrous texture and hence the modulus [53–55]. Due to its higher specific area, the presence of TO-NFC will provide attaching sites to facilitate cell attachment.

4.1 Optimizing Bio-inks and Precursors

By combining alginate, CMC, and TO-NFC as a hybrid hydrogel, we design to overcome their individual challenges as bio-inks. Incorporating CMC will help cell migration, which is often the weakness of alginate matrix materials. Additionally, with CMC being a viscosity modifier, a higher modulus can be achieved with less solid content in the bio-ink composition. Adding the high-aspect ratio of the nanofiber into the alginate matrix will increase its fibrous texture and hence the modulus. Due to its higher specific area, the presence of TO-NFC will provide an attaching site to facilitate cell attachment. Thus, we hypothesized that the proposed hybrid hydrogel composition (alginate, CMC, and TO-NFC) will demonstrate higher cell viability by facilitating cell migration and attachment due to the presence of micro-porosity inside the scaffold microenvironment. A comprehensive analysis includes rheological tests to evaluate the flow behavior and viscoelastic properties, which are crucial for assessing printability and material stability during extrusion-based fabrication. Mechanical testing is performed to measure the strength, elasticity, and durability of the hydrogels, ensuring their suitability for load-bearing applications or structural integrity in 3D scaffolds. The following sections discuss these physical characterization techniques in detail and their relevance to optimizing hybrid hydrogel formulations. The letters “A,” “C,” and “T” denote sodium alginate, carboxymethyl cellulose, and tempo-mediated nano-fibrillated cellulose, respectively.

Shear-Thinning Behavior: Shear thinning, also known as pseudoplastic behavior, is a rheological property in which the viscosity of a material decreases as the applied shear rate increases. In the context of hydrogels for 3D bioprinting, this behavior is critical for enabling smooth extrusion through the printer’s nozzle. Under high shear conditions, such as during extrusion, the hydrogel becomes less viscous, allowing it to flow easily [56, 57]. Once the shear force is removed, the material rapidly regains its higher viscosity, maintaining the printed structure’s shape and

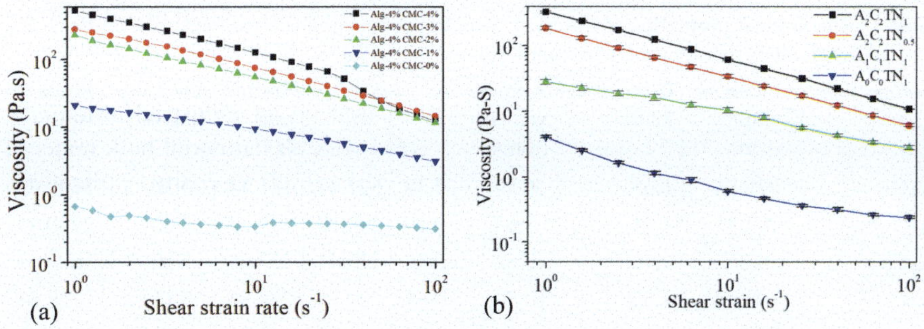

Fig. 11 Shear-thinning behavior of **a** alginate-CMC and **b** alginate-CMC-TO-NFC hydrogels at varying concentrations of components (log-log plot)

fidelity. This property is particularly beneficial in bioprinting applications as it minimizes the stress on encapsulated cells during deposition, enhances print resolution, and supports the creation of complex, stable 3D scaffolds. The ability to control shear-thinning behavior by modifying the hydrogel composition (e.g., adjusting polymer concentrations) ensures compatibility with specific bioprinting processes and desired scaffold properties [58].

Figure 11 demonstrates the shear-thinning behavior of various compositions under different shear strain rates varying from 1 to 100 s^{-1}. Viscosity decreases as the shear strain rate increases, indicating non-Newtonian behavior typical of shear-thinning materials. Additionally, using the following power-law equation, the compositions' flow characteristic parameters can be determined.

$$\eta = K\dot{\gamma}^{n-1} \tag{10}$$

where η is the viscosity, $\dot{\gamma}$ is the shear rate, and K and n are the shear-thinning coefficients.

Tables 1 and 2 provide the power-law parameters n (flow behavior index) and K (consistency index) for various compositions, calculated by fitting the power-law Eq. (10) to the viscosity-shear rate data as shown in Fig. 12a. The key observations for n and K are:

All values of n are less than 1, confirming that the hybrid hydrogels exhibit shear-thinning behavior.

Lower n values indicate stronger shear-thinning behavior. The crosslinking properties of the alginate material is disrupted by the presence of CMC and TO-NFC fibrils. As a result, the n value decreases, reflecting a more pronounced reduction in viscosity with increasing shear rate. This is consistent with the graph, where higher CMC content results in steeper viscosity reductions.

The K values indicate the initial viscosity of the hydrogels at low shear rates (~1). Higher K values correspond to higher material content, as seen in the graph where higher CMC content shows higher viscosities.

Table 1 Rheological properties of different A and C composition. 8 psi extrusion pressure and 410 μm nozzle diameter are used when appropriate

Compositions	Filament width (μm)	Correction factor	Average velocity (mm/s)	n	K	Shear rate (s^{-1})	Viscosity (Pa.S)	Shear stress (Pa)
A_4C_4	675.33	1.65	6.59	0.38	522.47	181.94	20.30	3692.83
A_4C_3	782.33	1.91	7.63	0.49	280.08	198.63	13.60	2700.93
A_4C_2	883.67	2.16	8.62	0.38	229.41	237.65	7.60	1806.61
A_4C_1	1086	2.65	10.59	0.65	20.95	234.53	3.11	728.64
A_4C_0	1280.67	3.12	12.49	0.70	0.67	269.76	0.13	34.19

Table 2 Rheological properties of different A, C, and T compositions. 10 psi extrusion pressure and 250 μm nozzle diameter are used when appropriate

Compositions	Filament width (μm)	Correction factor	Material velocity (mm/s)	n	K ($Pa.S^n$)	Shear rate (s^{-1})	Viscosity (Pa.S)
$A_2C_2T_1$	318	1.27	6.36	0.26	329	348.33	4.33
$A_2C_2T_{0.5}$	385	1.54	7.7	0.27	190	412.95	2.34
$A_1C_1T_1$	440	1.76	8.8	0.48	28	357.87	1.32
$A_0C_0T_1$	520	2.08	10.4	0.55	4	400.87	0.27

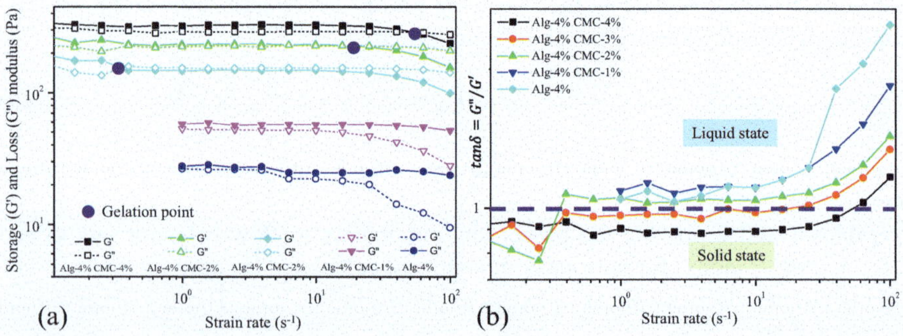

Fig. 12 Test results for storage modulus (G′), loss modulus (G″) for **a** A and C composition hybrid hydrogels **b** showing the tangent (*tanδ*) as functions of strain rate (%)

The correction factor is the ratio between extruded width and diameter and is measured after 2 min. A lower correction factor implies that the material after extrusion can hold the bond better, which can be interpreted as better mechanical properties.

Loss and Storage Modulus for Loss Tangent: The rheological behavior of hydrogels is often characterized by key parameters such as storage modulus (G′), loss modulus (G‴), linear viscoelastic range (LVR), and loss tangent (*tanδ*). These parameters provide insights into the material's ability to balance elastic and viscous properties under deformation, critical for applications like 3D bioprinting.

Storage Modulus (G′): The storage modulus represents the elastic component of a material's behavior, indicating the energy stored and recovered per cycle of deformation. A high G′ value suggests solid-like behavior, where the material resists flow and retains its structure.

Loss Modulus (G″): The loss modulus reflects the viscous component, representing the energy dissipated as heat during deformation. A dominant G″ value corresponds to liquid-like behavior, where the material flows more easily.

Linear Viscoelastic Range (LVR): The LVR defines the region where the material's structure remains intact and its response to applied strain is linear. Within this range, the material can recover its original structure upon the removal of stress without permanent deformation. The end of the LVR is typically identified by the intersection of G′ and G″, signaling the material's transition to flow-dominated behavior.

Loss Tangent (*tanδ*): The loss tangent is the ratio of G″ to G′ ($tan\delta = G''/G'$), quantifying the relative dominance of viscous or elastic behavior. A *tanδ* value less than 1 indicates predominantly elastic (solid-like) behavior, while a value greater than 1 indicates predominantly viscous (liquid-like) behavior.

The amplitude sweep test for the hybrid hydrogel compositions, conducted at 1 Hz, is illustrated in Figs. 12 and 13, showing the variation of storage modulus (G′) and loss modulus (G″) as a function of shear strain (%). The results reveal that increasing the percentage of CMC transforms the hydrogel's behavior from a liquid-like state to a solid-like state. Also, the LVR for only TO-NFC material can't be found between 1 and 100 shear rate, which implies its non-printability characters. By increasing the solid content in both compositions, the LVR moves to the right, meaning higher pressure is required to extrude the material. This transition is critical for understanding the relationship between extrusion pressure and material flow in 3D bioprinting. Effective extrusion pressure must exceed the LVR strain rate to ensure consistent material flow through the nozzle.

Hysteresis Loop and Recovery Rate: To understand the change in the gelation process during the extrusion of the hydrogel, the recovery rate under different applied pressure is a critical parameter. A stress is applied on the at-rest hydrogel, which breaks down the initial network structures of the hydrogel. After the hydrogel is extruded through the nozzle, due to the shear force, the bond between the polymer chain is disrupted. After the extrusion, the pressure is released, and the internal network bond starts recovering. In most cases, when the shear is released, the stress path lags, forming a hysteresis loop as shown in Fig. 14. The area within the hysteresis loop signifies the energy consumed in the internal network breakdown. This test is particularly relevant for 3D bioprinting, as it simulates the behavior of hydrogels during different phases of the extrusion process. The graph shows that pure Alginate is highly viscoelastic compared to the composition with C and T. An increase in C and T content signifies irreparable polymeric bond breakage under pressure. Additionally, the pure TO-NFC material shows minimal change in shear stress under different shear rates, which implies that no or minimal shear stress is required to induce flow and the material itself is not suitable for 3D bioprinting.

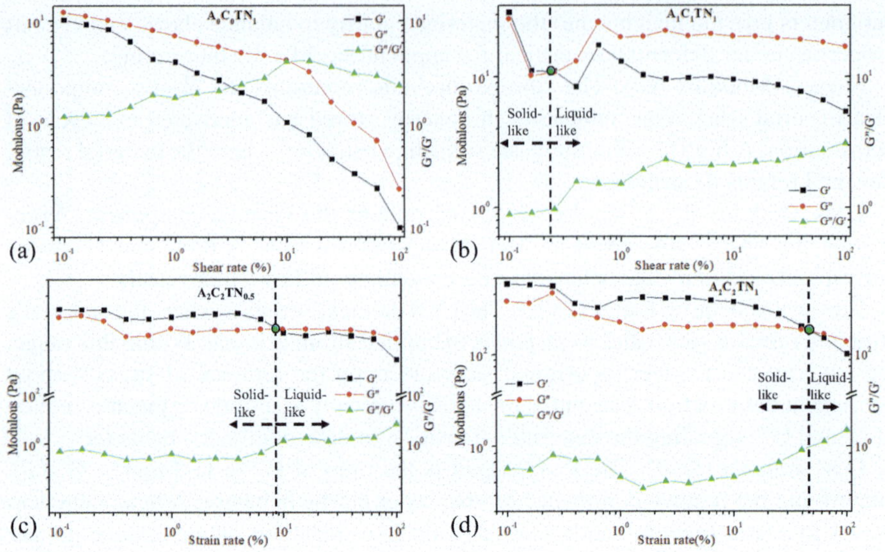

Fig. 13 Test results for storage modulus (G′), loss modulus (G′′) and tangent (*tanδ*) for A, C and T composition hybrid hydrogels

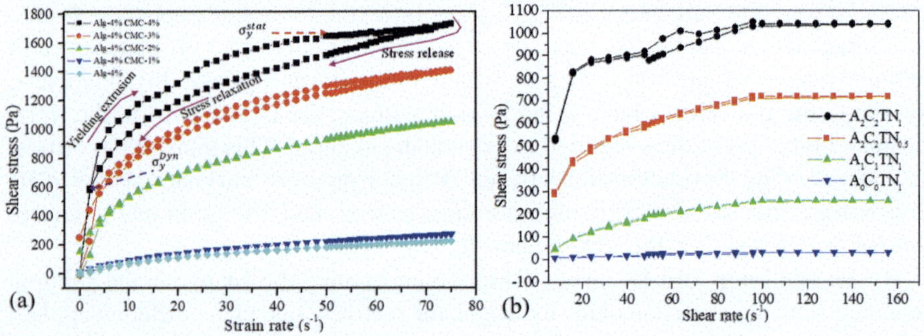

Fig. 14 Comparison of hysteresis loop for **a** A and C compositions and **b** A, C, and T compositions

5 3D Bioprinting

Inkjet, extrusion-based, and laser-assisted bioprinting systems are the most common bioprinting modalities. Depending upon the specific bioprinting process and technique employed, living cells undergo various challenges and stresses at each stage of bioprinting. For example, high shear stress can rupture the cell-wall during the extrusion process at the nozzle tip. Similarly, the absorbed energy (e.g., laser or heat) can alter the DNA or the phenotype of living cells and undermine the purpose of bioprinting. Developed in the late 80 s' as "Cytoscribing," this extrusion-based

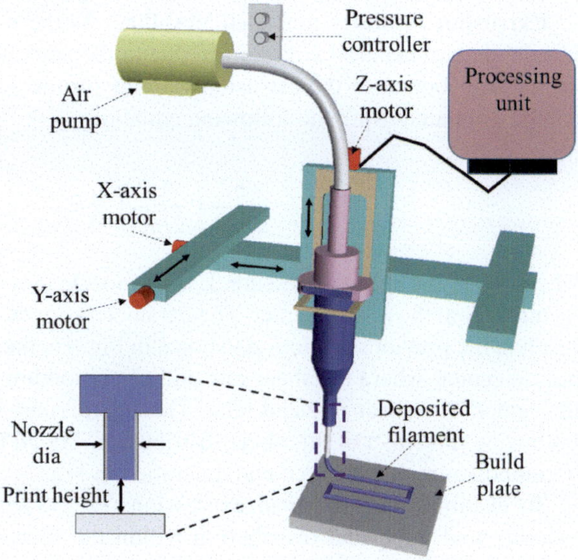

Fig. 15 Schematic diagram of a 3D bioprinter

bioprinting is a relatively simple cell writing process that can provide a sterilized environment during fabrication. It is also capable of delivering multiple materials and heterogeneous cell type combinations through a closed-loop system while maintaining their phenotype with appropriately designed shear-thinning materials. Both acellular and cell-encapsulated biomaterials can be extruded by controlling the printing parameters. Thus, extrusion-based bioprinting is the focus of this chapter. The schematic (Fig. 15) illustrates the core components and operational principles of a typical extrusion-based 3D bioprinter. The system comprises:

Processing Unit: A computer-controlled unit that guides the printing process, including material deposition paths and printing parameters.

Pressure Controller: Regulates the extrusion pressure applied to the bio-ink, ensuring consistent flow through the nozzle.

Motors (X, Y, Z Axes): Enable precise movement of the print head or build plate along the three axes, allowing accurate layer-by-layer material deposition.

Extrusion Nozzle: Deposits the bio-ink in the desired pattern. The nozzle's geometry, diameter, and height above the build plate significantly affect the filament's shape and fidelity.

Build Plate: Serves as the platform for the printed structure, providing stability, and precision during fabrication.

The zoomed section highlights the deposition of a filament on the build plate, emphasizing the importance of maintaining accurate nozzle height and extrusion flow for achieving uniform layers. This configuration ensures that the bio-ink is deposited with high precision, enabling the creation of 3D scaffolds with complex architectures for applications such as tissue engineering and regenerative medicine.

Extrusion Pressure and Cell Viability: As the material is pushed through the nozzle, the gel material will deform due to the applied shear and start flowing. The spatial distribution of the extrusion velocity profile across the nozzle cross-section can be estimated using the following equation:

$$v = \frac{n}{n+1}\left(\frac{\Delta p}{2LK}\right)^{\frac{1}{n}}\left(R^{\frac{n+1}{n}} - r^{\frac{n+1}{n}}\right) \tag{11}$$

where Δp is the applied pressure, L is the nozzle length, and r is the distance of any location from the nozzle center. The parameters are determined from the rheological result in the previous section. As shown in Fig. 16, the velocity increases toward the nozzle center, where the shear rate and corresponding shear stress decrease toward the center. The shear rate and related shear stress are maximum at the nozzle wall. Following the effect of this shear distribution, TO-NFC fiber alignment in A, C and T compositions is measured and quantified in Fig. 17.

To examine the analytical expression (Eq. 11) of extrusion pressure on cell viability and its spatial distribution within the filament, HEK 293 cells encapsulated in an A_4C_4 hydrogel composition were printed at 8 psi and 12 psi. Shear stress calculations indicated that cells near the nozzle wall would experience higher shear

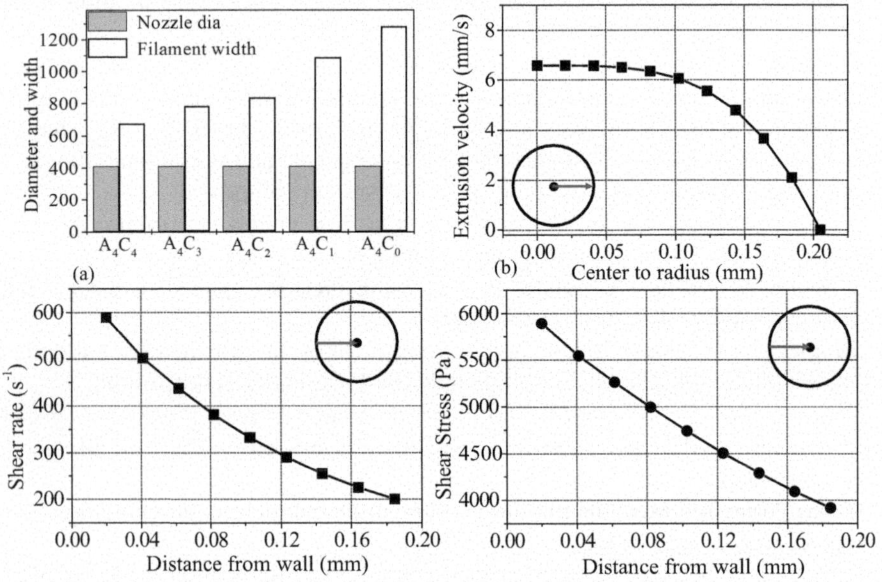

Fig. 16 **a** Filament width of various compositions, **b** distribution of velocity from the center to the radius, **c** distribution of shear rate from the nozzle wall toward the center, and **d** distribution of corresponding shear stress from the nozzle wall toward the center. 410-micron nozzle and A-C compositions are used in these calculations

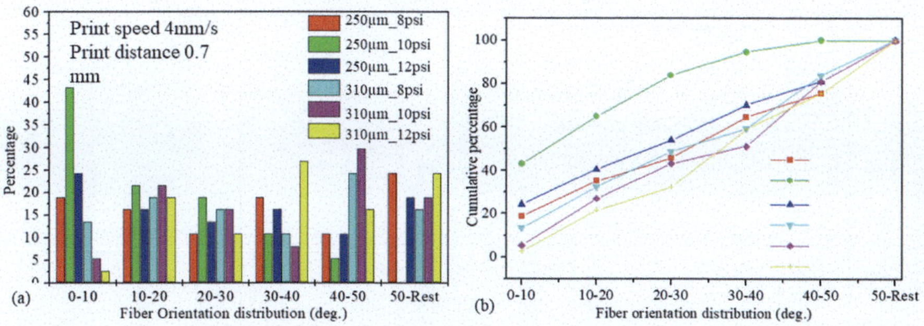

Fig. 17 Influence of nozzle diameter and extrusion pressure on fiber alignment **a** percentage distribution and **b** cumulative distribution using $A_2C_2T_1$ Composition

stress, potentially leading to lower cell viability at the filament's edge as shown in Fig. 18.

Mechanical Integrity Test: The filament collapse test evaluates the mid-span deflection of a suspended filament by analyzing its ability to form a bridge between two pillars. A platform consisting of seven pillars with known spacing (1 through 6 mm) is fabricated using a 3D printing process to ensure consistent geometry, as shown in Fig. 19. A single filament of the composition is deposited on the platform, spanned across the pillars. The collapse area factor is determined as the percentage of the actual area of the deflected filament relative to the theoretical area, which demonstrates the self-supporting mechanical integrity of the composition (cf. Fig. 19). A smaller value represents a better mechanical property. The addition of TO-NFC in the composition acts as a reinforcing fiber which enhances the mechanical property and as a result shows a better performance even with lower solid content (Table 3).

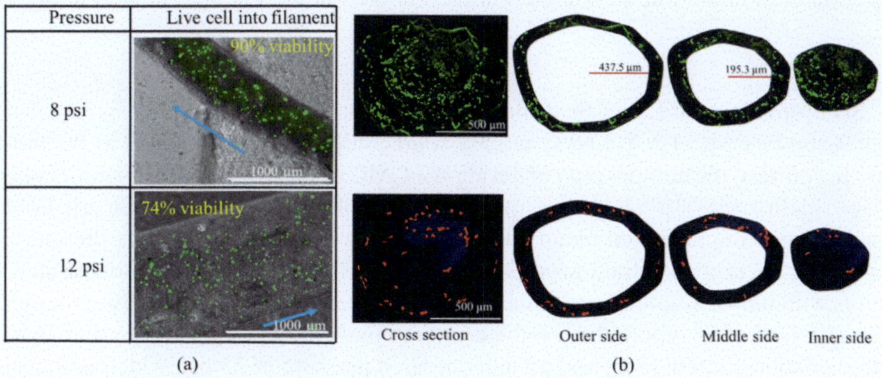

Fig. 18 **a** Cell viability in 4% alginate-4%CMC scaffolds over time, showing 90% versus 74% viability at two different applied pressures and **b** Comparison of the presence of live/dead cell on the cross-section of a filament deposited at 8 psi

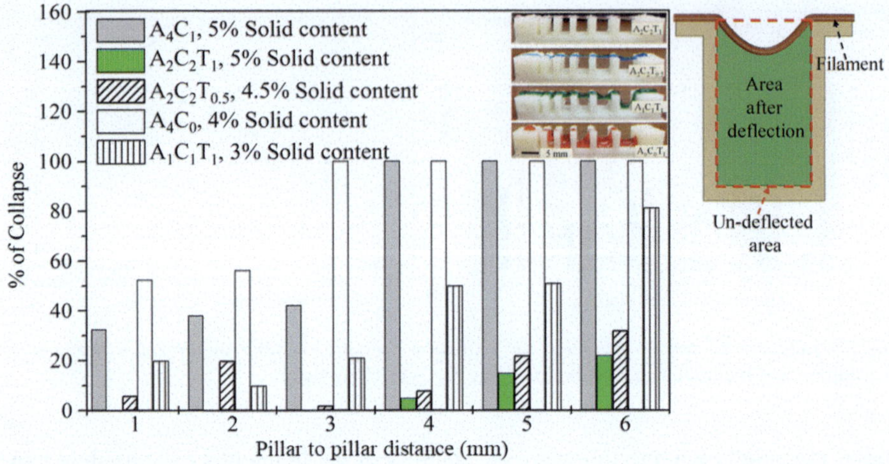

Fig. 19 Comparison of the percentage of collapse with and without TO-NFC into alginate and TO-NFC at different percentages of solid content

Table 3 Comparison of the collapse area factor percentage for A-C and A-C-T compositions

	Alginate and CMC					Alginate, CMC and TO-NFC			
Pillar distance	A_4C_4	A_4C_3	A_4C_2	A_4C_1	A_4C_0	$A_2C_2T_1$	$A_2C_2T_{0.5}$	$A_1C_1T_1$	$A_0C_0T_1$
1	0	0	0	32.5	52.5	0	6	20	100
2	0	0	5	38	56.25	0	20	10	100
3	0	5	15	42.25	100	0	2	21	100
4	2.5	5.63	17.5	100	100	5	8	50	100
5	4.5	8.85	28.85	100	100	15	22	51	100
6	4.58	16.13	32.38	100	100	22	32	81	100

3D Bioprinting of Large Scaffolds In post-characterization analysis, it is revealed that %alg-2%CMC-1% TO-NFC at a 5% solid content can perform well as the bio-ink. In contrast, the composition of 4%alg-4%CMC also shows good printability and shape fidelity with 8% solid content (Fig. 20). Table 4 represents a comparison of data between the proposed blend and other bio-ink presented in various literature. Following the proposed framework, our hybrid hydrogel is capable to fabricate more layers with only 5% solid content and relatively lower applied pressure. Even though some literature has reported the same or more numbers of printed layers, they used a higher solid content (8%) and a higher applied pressure of 35 psi, which both can be detrimental to the living cells.

Cell Viability: The cell viability and biocompatibility of encapsulated Porc 1 cells in a 2%alg-2%CMC-1% TO-NFC composition are assessed using a LIVE/

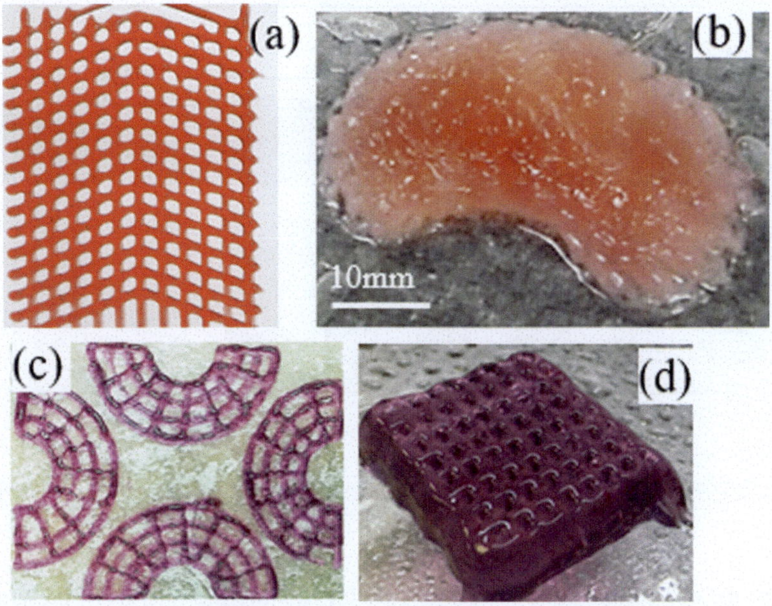

Fig. 20 Large-scale scaffold fabricated by **a–b** 4%alg-4%CMC, **c–d** 2%alg-2%CMC-1% TO-NFC

Table 4 Comparison between our proposed bio-ink to existing bio-inks

Solid content (%)	Components	Viscosity range (Pa.S)	Number of layers printed	Filament fusion rate (%)	Applied pressure (psi)	Ref
12 (3/9)	Alg/MC	800–30	50	3	58	[11]
8	MC	500–30	42	148	35	[59]
4	Potato starch	500–40	8	NA	NA	[60]
9(3/3/3)	Lap/alg/MC	1000–100	30	6	11.6–13.78	[61]
6.7(2.8/0.9/3)	Alg, aga, MC	580–10	20	2.13	11–14	[62]
3	hdECM	45–0.9	10	73	NA	[63]
9	Alg/$CaCl_2$	300–1	NA	157	29	[64]
5(2/2/1)	*Alg/CMC/TO-NFC*	*329–30*	*42*	*27*	10	*Proposed*

DEAD assay after printing and at different time intervals as shown in Fig. 21. Filaments fabricated with various air pressures, i.e., 8, 10, and 12 psi, show different cell viability. Their viability significantly increases (p-value = 0.0034 with 95% confidence interval) up to 90%, 89%, and 81%, respectively after 10 days of incubation, reflecting the proof of cell proliferation into the composition. The percentage of cell

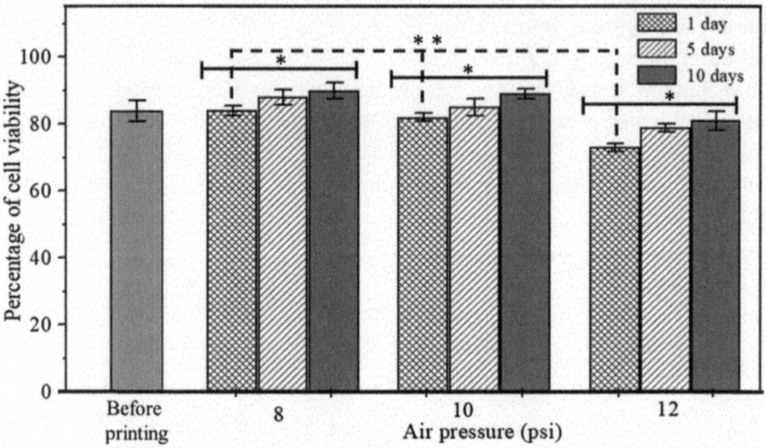

Fig. 21 Percentage of cell viability before printing and at different air pressures and incubation days

viability for our composition is compared with existing bio-inks from literature in Table 4.

Additionally, the performance of cellulose fiber for cell viability is also compared in Table 5. A material with a shorter fiber size represents a more fibrous texture into the composition, which creates a highly micro-level porous environment for the encapsulated cell. The scaffold environment having micro-porosity will allow the transportation of necessary growth media and gas. Moreover, the higher specific area due to the presence of nanofiber increases the attaching site significantly, which facilitates cell attachment. The microenvironment can partially mimic the topographical features of the natural ECM and influence cell activity [65]. Moreover, it is reported that the presence of an anionic group in NFC (e.g., TO-NFC) can result in the conformation of adsorbed protein. This scenario may direct the activation and adhesion of the encapsulated cell [66]. Hence, more cell clusters are visible for the composition of $A_2C_2T_1$ due to the increased cell-material and cell–cell interactions.

Table 5 Comparison of cell viability with cellulose-based bio-inks and a comparison with alginate

Bio-ink constituents	Solid content (%)	Incubation day	Cell viability (%)
Pure alginate [40]	3	21	80
Alg-laponite-methyl cellulose [61]	9	21	70
Alg-methyl cellulose [40]	12	21	78
Alg-CMC-MMT [61, 67]	10	7	84
Alg-CMC [19]	8	23	86
Alg-CMC-TO-NFC [31]	5	10	90

6 Conclusion

Successful bioprinting has a complex relationship between data, material-process and functions. Although numerous studies have demonstrated cellulose as an effective biomaterial, focusing only on one aspect may not be enough for its successful implementation in this domain. It is important to maintain a seamless transition between each aspect of the bioprinting ecosystem (from digital to physical). Current research practice for optimizing bio-inks involves exhaustive experimentation with multi-material compositions. Predicting bio-ink properties can be beneficial to the research community but is a challenging task due to their non-Newtonian behavior in complex compositions. Data driven machine learning frameworks have become a new but popular tool to predict the composition of bio-inks and their performance, which can be the future direction of bioprinting research.

References

1. Agarwal, S., Saha, S., Balla, V.K., Pal, A., Barui, A., Bodhak, S.: Current developments in 3D bioprinting for tissue and organ regeneration—a review. Front. Mech. Eng. **6** (2020). https://doi.org/10.3389/fmech.2020.589171
2. Dababneh, A.B., Ozbolat, I.T.: Bioprinting technology: a current state-of-the-art review. J. Manuf. Sci. Eng. **136** (2014)
3. Murphy, S.V., Atala, A.: 3D bioprinting of tissues and organs. Nat. Biotechnol. **32**, 773–785 (2014)
4. Jia, W., Gungor-Ozkerim, P.S., Zhang, Y.S., Yue, K., Zhu, K., Liu, W., et al.: Direct 3D bioprinting of perfusable vascular constructs using a blend bioink. Biomaterials **106**, 58–68 (2016)
5. Axpe, E., Oyen, M.L.: Applications of alginate-based bioinks in 3D bioprinting. Int. J. Mol. Sci. **17**, 1976 (2016)
6. Skardal, A., Atala, A.: Biomaterials for integration with 3-D bioprinting. Ann. Biomed. Eng. **43**, 730–746 (2015)
7. Yeong, W.-Y., Chua, C.-K., Leong, K.-F., Chandrasekaran, M.: Rapid prototyping in tissue engineering: challenges and potential. Trends Biotechnol. **22**, 643–652 (2004)
8. Leong, K., Cheah, C., Chua, C.: Solid freeform fabrication of three-dimensional scaffolds for engineering replacement tissues and organs. Biomaterials **24**, 2363–2378 (2003)
9. Pham, Q.P., Sharma, U., Mikos, A.G.: Electrospinning of polymeric nanofibers for tissue engineering applications: a review. Tissue Eng. **12**, 1197–1211 (2006)
10. Vijayavenkataraman, S., Shuo, Z., Fuh, J.Y., Lu, W.F.: Design of three-dimensional scaffolds with tunable matrix stiffness for directing stem cell lineage specification: an in silico study. Bioengineering **4**, 66 (2017)
11. Li, H., Tan, Y.J., Leong, K.F., Li, L.: 3D bioprinting of highly thixotropic alginate/methylcellulose hydrogel with strong interface bonding. ACS Appl. Mater. Interfaces **9**, 20086–20097 (2017)
12. Kulkarni, P., Marsan, A., Dutta, D.: A review of process planning techniques in layered manufacturing. Rapid Prototyp. J. **6**, 18–35 (2000)
13. Khoda, A.K.M.B., Ozbolat, I.T., Koc, B.: Engineered tissue scaffolds with variational porous architecture. J. Biomech. Eng. **133**, 011001 (2011)
14. Deo, K.A., Singh, K.A., Peak, C.W., Alge, D.L., Gaharwar, A.K.: Bioprinting 101: design, fabrication, and evaluation of cell-laden 3D bioprinted scaffolds. Tissue Eng. Part A **26**, 318–338 (2020)

15. Gungor-Ozkerim, P.S., Inci, I., Zhang, Y.S., Khademhosseini, A., Dokmeci, M.R.: Bioinks for 3D bioprinting: an overview. Biomater. Sci. **6**, 915–946 (2018)
16. Cipitria, A., Lange, C., Schell, H., Wagermaier, W., Reichert, J.C., Hutmacher, D.W., et al.: Porous scaffold architecture guides tissue formation. J. Bone Miner. Res. **27**, 1275–1288 (2012)
17. Kang, H.-W., Lee, S.J., Ko, I.K., Kengla, C., Yoo, J.J., Atala, A.: A 3D bioprinting system to produce human-scale tissue constructs with structural integrity. Nat. Biotechnol. **34**, 312–319 (2016)
18. Habib, M.A., Khoda, B.: Attribute driven process architecture for additive manufacturing. Robot. Comput. Integr. Manuf. **44**, 253–265 (2017)
19. Habib, A., Sathish, V., Mallik, S., Khoda, B.: 3D printability of alginate-carboxymethyl cellulose hydrogel. Materials **11**, 454 (2018)
20. Ahsan, N., Khoda, B.: AM optimization framework for part and process attributes through geometric analysis. Addit. Manuf. **11**, 85–96 (2016). https://doi.org/10.1016/j.addma.2016.05.013
21. Murphy, C.M., Haugh, M.G., O'Brien, F.J.: The effect of mean pore size on cell attachment, proliferation and migration in collagen–glycosaminoglycan scaffolds for bone tissue engineering. Biomaterials **31**, 461–466 (2010)
22. Perez, R.A., Mestres, G.: Role of pore size and morphology in musculo-skeletal tissue regeneration. Mater. Sci. Eng. C **61**, 922–939 (2016)
23. Byun, H.-S., Lee, K.H.: Determination of the optimal build direction for different rapid prototyping processes using multi-criterion decision making. Robot. Comput. Integr. Manuf. **22**, 69–80 (2006)
24. Frank, D., Fadel, G.: Expert system-based selection of the preferred direction of build for rapid prototyping processes. J. Intell. Manuf. **6**, 339–345 (1995)
25. See, T.-K., Gurnani, A., Lewis, K.: Multi-attribute decision making using hypothetical equivalents and inequivalents. J. Mech. Des. **126**, 950–958 (2004). https://doi.org/10.1115/1.1814389
26. Floyd, R.W., Steinberg, L.: An adaptive algorithm for spatial grey scale. Soc. Inf. Display **17**, 75–77 (1976)
27. Izadifar, Z., Chen, X., Kulyk, W.: Strategic design and fabrication of engineered scaffolds for articular cartilage repair. J. Funct. Biomater. **3**, 799–838 (2012). https://doi.org/10.3390/jfb3040799
28. Tang, D., Tare, R.S., Yang, L.-Y., Williams, D.F., Ou, K.-L., Oreffo, R.O.C.: Bio-fabrication of bone tissue: approaches, challenges and translation for bone regeneration. Biomaterials **83**, 363–382 (2016)
29. Wang, X., Xu, S., Zhou, S., Xu, W., Leary, M., Choong, P., et al.: Topological design and additive manufacturing of porous metals for bone scaffolds and orthopaedic implants: a review. Biomaterials **83**, 127–141 (2016). https://doi.org/10.1016/j.biomaterials.2016.01.012
30. Ahsan, A.N., Xie, R., Khoda, B.: Direct bio-printing with heterogeneous topology design. In: Proceeding SME NAMRC 45, Los Angeles, CA, USA (2017)
31. Habib, M., Khoda, B.: Fiber filled hybrid hydrogel for bio-manufacturing. J. Manuf. Sci. Eng. 1–38 (2020)
32. Gasperini, L., Mano, J.F., Reis, R.L.: Natural polymers for the microencapsulation of cells. J. R. Soc. Interface. **11**, 20140817 (2014)
33. Hunt, N.C., Shelton, R.M., Henderson, D.J., Grover, L.M.: Calcium-alginate hydrogel-encapsulated fibroblasts provide sustained release of vascular endothelial growth factor. Tissue Eng. Part A **19**, 905–914 (2012)
34. Sapir, Y., Cohen, S., Friedman, G., Polyak, B.: The promotion of in vitro vessel-like organization of endothelial cells in magnetically responsive alginate scaffolds. Biomaterials **33**, 4100–4109 (2012)
35. Kundu, J., Shim, J.H., Jang, J., Kim, S.W., Cho, D.W.: An additive manufacturing-based PCL–alginate–chondrocyte bioprinted scaffold for cartilage tissue engineering. J. Tissue Eng. Regen. Med. **9**, 1286–1297 (2015)

36. Mosahebi, A., Simon, M., Wiberg, M., Terenghi, G.: A novel use of alginate hydrogel as Schwann cell matrix. Tissue Eng. **7**, 525–534 (2001)
37. Hunt, N., Smith, A.M., Gbureck, U., Shelton, R., Grover, L.: Encapsulation of fibroblasts causes accelerated alginate hydrogel degradation. Acta Biomater. **6**, 3649–3656 (2010)
38. Sun, J., Tan, H.: Alginate-based biomaterials for regenerative medicine applications. Materials **6**, 1285–1309 (2013)
39. Di Giuseppe, M., Law, N., Webb, B., Macrae, R.A., Liew, L.J., Sercombe, T.B., et al.: Mechanical behaviour of alginate-gelatin hydrogels for 3D bioprinting. J. Mech. Behav. Biomed. Mater. **79**, 150–182 (2018)
40. Schütz, K., Placht, A.M., Paul, B., Brüggemeier, S., Gelinsky, M., Lode, A.: Three-dimensional plotting of a cell-laden alginate/methylcellulose blend: towards biofabrication of tissue engineering constructs with clinically relevant dimensions. J. Tissue Eng. Regen. Med. **11**, 1574–1587 (2017)
41. Markstedt, K., Mantas, A., Tournier, I., Martínez Ávila, H.C., Hägg, D., Gatenholm, P.: 3D bioprinting human chondrocytes with nanocellulose–alginate bioink for cartilage tissue engineering applications. Biomacromolecules **16**, 1489–1496 (2015)
42. Han, Y., Wang, L.: Sodium alginate/carboxymethyl cellulose films containing pyrogallol acid: physical and antibacterial properties. J. Sci. Food Agric. **97**, 1295–1301 (2017)
43. Garrett, Q., Simmons, P.A., Xu, S., Vehige, J., Zhao, Z., Ehrmann, K., Willcox, M.: Carboxymethylcellulose binds to human corneal epithelial cells and is a modulator of corneal epithelial wound healing. Investig. Ophthalmol. Vis. Sci. **48**, 1559–1567 (2007)
44. Kim, M.S., Park, S.J., Gu, B.K., Kim, C.-H.: Ionically crosslinked alginate–carboxymethyl cellulose beads for the delivery of protein therapeutics. Appl. Surf. Sci. **262**, 28–33 (2012)
45. Agarwal, T., Narayana, S.G.H., Pal, K., Pramanik, K., Giri, S., Banerjee, I.: Calcium alginate-carboxymethyl cellulose beads for colon-targeted drug delivery. Int. J. Biol. Macromol. **75**, 409–417 (2015)
46. Fukuzumi, H., Tanaka, R., Saito, T., Isogai, A.: Dispersion stability and aggregation behavior of TEMPO-oxidized cellulose nanofibrils in water as a function of salt addition. Cellulose **21**, 1553–1559 (2014)
47. Abe, K., Yano, H.: Cellulose nanofiber-based hydrogels with high mechanical strength. Cellulose **19**, 1907–1912 (2012)
48. Roman, M.: Toxicity of cellulose nanocrystals: a review. Ind. Biotechnol. **11**, 25–33 (2015)
49. Siró, I., Plackett, D.: Microfibrillated cellulose and new nanocomposite materials: a review. Cellulose **17**, 459–494 (2010)
50. Narayanan, L.K., Huebner, P., Fisher, M.B., Spang, J.T., Starly, B., Shirwaiker, R.A.: 3D-bioprinting of polylactic acid (PLA) nanofiber–alginate hydrogel bioink containing human adipose-derived stem cells. ACS Biomater. Sci. Eng. **2**, 1732–1742 (2016)
51. Nguyen, D., Hägg, D.A., Forsman, A., Ekholm, J., Nimkingratana, P., Brantsing, C., et al.: Cartilage tissue engineering by the 3D bioprinting of iPS cells in a nano-cellulose/alginate bio-ink. Sci. Rep. **7**, 658 (2017)
52. Li, V.C., Mulyadi, A., Dunn, C.K., Deng, Y., Qi, H.J.: Direct ink write 3D printed cellulose nanofiber aerogel structures with highly deformable, shape recoverable, and functionalizable properties. ACS Sustain. Chem. Eng. **6**, 2011–2022 (2018)
53. Torres-Rendon, J.G., Köpf, M., Gehlen, D., Blaeser, A., Fischer, H., Laporte, L.D., Walther, A.: Cellulose nanofibril hydrogel tubes as sacrificial templates for freestanding tubular cell constructs. Biomacromol **17**, 905–913 (2016)
54. Lin, N., Dufresne, A.: Nanocellulose in biomedicine: current status and future prospect. Eur. Polym. J. **59**, 302–325 (2014)
55. Mohammadi, P., Toivonen, M.S., Ikkala, O., Wagermaier, W., Linder, M.B.: Aligning cellulose nanofibril dispersions for tougher fibers. Sci. Rep. **7**, 1–10 (2017)
56. Diamantides, N., Wang, L., Pruiksma, T., Siemiatkoski, J., Dugopolski, C., Shortkroff, S., et al.: Correlating rheological properties and printability of collagen bioinks: the effects of riboflavin photocrosslinking and pH. Biofabrication **9**, 034102 (2017)

57. Jiang, Y., Zhou, J., Feng, C., Shi, H., Zhao, G., Bian, Y.: Rheological behavior, 3D printability and the formation of scaffolds with cellulose nanocrystals/gelatin hydrogels. J. Mater. Sci. **55**, 15709–15725 (2020)
58. Habib, M.A., Khoda, B.: Rheological analysis of bio-ink for 3D bioprinting processes. J. Manuf. Process. **76**, 708–718 (2022). https://doi.org/10.1016/j.jmapro.2022.02.048
59. Ahlfeld, T., Köhler, T., Czichy, C., Lode, A., Gelinsky, M.: A methylcellulose hydrogel as support for 3D plotting of complex shaped calcium phosphate scaffolds. Gels **4**, 68 (2018)
60. Liu, Z., Zhang, M., Bhandari, B., Yang, C.: Impact of rheological properties of mashed potatoes on 3D printing. J. Food Eng. **220**, 76–82 (2018)
61. Ahlfeld, T., Cidonio, G., Kilian, D., Duin, S., Akkineni, A., Dawson, J., et al.: Development of a clay based bioink for 3D cell printing for skeletal application. Biofabrication **9**, 034103 (2017)
62. Seidel, J., Ahlfeld, T., Adolph, M., Kümmritz, S., Steingroewer, J., Krujatz, F., et al.: Green bioprinting: extrusion-based fabrication of plant cell-laden biopolymer hydrogel scaffolds. Biofabrication **9**, 045011 (2017)
63. Jang, J., Kim, T.G., Kim, B.S., Kim, S.-W., Kwon, S.-M., Cho, D.-W.: Tailoring mechanical properties of decellularized extracellular matrix bio-ink by vitamin B2-induced photocrosslinking. Acta Biomater. **33**, 88–95 (2016)
64. Paxton, N.C., Smolan, W., Böck, T., Melchels, F.P., Groll, J., Juengst, T.: Proposal to assess printability of bio-inks for extrusion-based bio-printing and evaluation of rheological properties governing bioprintability. Biofabrication (2017)
65. Guo, Z., Ma, M., Huang, X., Li, H., Zhou, C.: Effect of fiber diameter on proliferation and differentiation of MC3T3-E1 pre-osteoblasts. J. Biomater. Tissue Eng. **7**, 162–169 (2017)
66. Hua, K., Ålander, E., Lindström, T., Mihranyan, A., Strømme, M., Ferraz, N.: Surface chemistry of nanocellulose fibers directs monocyte/macrophage response. Biomacromol **16**, 2787–2795 (2015)
67. Habib, A., Khoda, B.: Development of clay based novel hybrid bio-ink for 3D bio-printing process. J. Manuf. Process. **38**, 76–87 (2019)

Bashir Khoda is an Associate Professor in the Department of Mechanical Engineering at the University at Maine. He received his Ph.D. in Industrial and Systems Engineering from the University at Buffalo (SUNY), New York. He leads research in digital manufacturing, with expertise in CAD/CAM/CAE, advanced manufacturing, 3D printing, and biomanufacturing. His work focuses on developing computational design tools and advanced fabrication methods for engineering applications ranging from smart materials to biomedical devices. Dr. Khoda has published extensively in these fields and actively contributes to interdisciplinary research initiatives. He is committed to advancing scalable and intelligent manufacturing solutions through academic and industry collaborations.

Md. Ahasan Habib is an Assistant Professor in the Department of Manufacturing and Mechanical Engineering Technology at Rochester Institute of Technology (RIT). He received his Ph.D. in Industrial and Manufacturing Engineering from North Dakota State University. His research focuses on robot-assisted additive biomanufacturing and the design and development of novel bioinks for tissue engineering applications. Dr. Habib integrates multi-material 3D printing, cell-laden hydrogel systems, and smart manufacturing approaches to fabricate high-fidelity, functional constructs. His work aims to bridge engineering innovations with biomedical needs. He actively mentors diverse student researchers and contributes to interdisciplinary collaborations.

Chapter 6
Living Composites: Mycelium Biofabrication and the Evolution of Design

David Andréen

Abstract This chapter examines the integration of living materials into architectural design, focusing on the use of mycelium composites formed through fungal biotransformation. As architecture faces increasing demands for environmental responsibility, material innovation has turned toward biological systems capable of replacing high-impact construction materials. Mycelium, the filamentous body of fungi, offers a biologically active and biodegradable alternative. When cultivated on lignocellulosic waste substrates, it produces composites that are lightweight, renewable, and suitable for a wide range of applications. The chapter provides an overview of the historical development, material behavior, and fabrication strategies associated with mycelium-based construction. It highlights how advances in additive manufacturing have enabled complex forms that support both fungal viability and architectural function. By combining digital design with microbial growth, new techniques allow for the creation of geometries that optimize aeration, structural performance, and environmental interaction. Central to the discussion is the concept of transcalar design: Biological processes at the molecular and cellular level are shown to influence, and be influenced by, decisions at the component and building scale. Computational tools support this multi-scalar coordination by enabling responsive geometries and adaptive material behavior. The chapter argues for a shift in architectural thinking toward systems that incorporate growth, decay, and regeneration as intrinsic design parameters. By rethinking materiality through the lens of living systems, the chapter proposes a framework for sustainable construction rooted in biological intelligence. This approach challenges traditional notions of permanence and uniformity, offering new possibilities for ecological integration and design agency.

Keywords Mycelium composites · Biodesign · Transcalar design · Additive manufacturing · Living materials · Sustainable architecture · Circular construction · Computational design in architecture

---

D. Andréen (✉)
bioDigital matter, Department of Architecture and Built Environment, Lund University, Lund, Sweden
e-mail: david.andreen@abm.lth.se

© The Author(s), under exclusive license to Springer Nature Singapore Pte Ltd. 2026

C. Klemmt and G. Beaucage (eds.), *Biodegradable 3D-Printing*, The Materials Research Society Series, https://doi.org/10.1007/978-981-95-4025-9_6

1 Introduction

The construction industry faces mounting pressure to evolve toward more sustainable, circular, and environmentally conscious practices. As global urban populations rise and climate challenges intensify, architecture plays an increasingly pivotal role in shaping not only social and cultural landscapes, but also our planet's ecological balance. Traditional construction materials such as concrete, steel, and petrochemical-based plastics have been prevalent for decades due to their reliability, availability, and engineering performance. Yet these materials often incur high environmental costs, including carbon-intensive production processes, non-renewable resource depletion, and challenges in recycling or disposal.

In response, research into "biodesign"—that is, the integration of living or biologically derived materials into architectural design—has gained significant momentum. Fungal biotransformation represents one of the most promising areas within this emerging paradigm. By harnessing the growth and bonding properties of fungal mycelium, designers and researchers have begun to develop new classes of construction materials that are lightweight, biodegradable, potentially carbon-sequestering, and conducive to cradle-to-cradle life cycles. In parallel, advancements in additive fabrication have made it possible to shape these biomaterials into complex forms that were previously difficult, if not impossible, to achieve through traditional casting or molding alone.

This chapter explores how fungal biotransformation can potentially redefine architectural evolution when coupled with additive fabrication (3D printing). Rather than providing highly specialized technical details, it offers a broader overview of the possibilities, challenges, and general trends for integrating living processes into the design and production of buildings. Readers will find discussions spanning from the nanoscale transformations that fungi perform on organic material, to the architectural scale of entire buildings. Along the way, we will highlight how computational design methodologies, material science innovations, and transdisciplinary collaboration all converge to enable more resilient, adaptive, and sustainable architectural solutions.

The chapter draws on multiple case studies from a variety of labs and industrial collaborations, but foremost that of our own group, bioDigital matter. By focusing on fungal-lignocellulosic composites and their potential for 3D printing, we situate this research in the broader domain of biodegradable and bio-based additive manufacturing. Through an emphasis on design strategies and general insights, our aim is to illuminate how fungal biotransformation can transform not only material choices, but also challenge deeper assumptions about the relationship between architects, buildings, and living organisms.

2 Background and Context

Biodesign, broadly defined, involves the incorporation of living or biologically derived materials in architectural practice. Historically, architecture has drawn aesthetic or structural inspiration from natural forms—exemplified by movements like Art Nouveau or the parametric biomorphism of the early digital design era. However, contemporary biodesign goes further by inviting organisms themselves to participate as co-creators, synthesizing new matter or adapting to local contexts in real time. This agency of the material itself is a powerful mechanism for addressing the complexity of transcalar design contexts, unleashing new strategies for architectural performance. The term "transcalar" refers to systems operating across multiple scales, such as spatial, temporal or organizational. In biohybrid or living materials, transcalar systems integrate living cells or organisms into engineered materials to create responsive, adaptive properties relevant across scales from the organisms to the city.

The rise of biodesign is fueled by an urgent search for more ecological alternatives. Construction has historically relied on linear consumption patterns: mine resources, process them, build, and eventually dispose. The results, from massive waste generation to excessive carbon footprints, have become untenable. By contrast, biodesign points toward circular and regenerative models: building materials grow from renewable or waste resources, have minimized net emissions, and ideally biodegrade or get repurposed at the end of their use cycle.

2.1 Historical Development of Mycelium Composites

Mycelium-based composites emerged in the late 2000s as a sustainable alternative to synthetics. The concept was pioneered by Eben Bayer and Gavin McIntyre, who in 2007 founded Ecovative Design after developing a "mushroom insulation" prototype in college (initially called Greensulate). By 2008–2010, they had patented the technology and commercialized Mushroom Packaging, a mycelium-grown foam marketed as EcoCradle, which gained early adoption by well-known brands as protective packaging. This period established the basic method of growing fungi on agricultural wastes to produce biodegradable composites and proved the concept's viability outside the lab.

Around the same time, artists and designers explored mycotecture, architectural applications of mycelium. Notably, artist Phil Ross cultivated fungus into furniture and small building components (coining the term "mycotecture") [1], and in 2014 architects demonstrated the first large-scale mycelium structure. The Hy-Fi tower by The Living, in collaboration with Ecovative and Arup, showcased mycelium blocks grown from agricultural waste and fungal strains, forming a temporary 12-meter-high tower in MoMA PS1's courtyard [2]. This pioneering project brought the potential of mycelium construction to the public mind. Since then, numerous approaches have been explored, including our own Protomycokion, which combines the raw potential of fungal materials with additive, digital fabrication processes that enable responsive, functionally graded, and articulated structures that unlock the opportunities inherent in a new building paradigm.

Following Hy-Fi, academic and commercial interest in mycelium composites accelerated. Research groups worldwide began experimenting with fungal building blocks. In 2017, engineers at Karlsruhe Institute (KIT) and ETH built MycoTree, an interlocking mycelium column exhibited at the Seoul Biennale, demonstrating that load-bearing spatial structures could be achieved by mycelium grown in molded units [3]. Meanwhile, startups expanded into new domains: MycoWorks, founded in 2013 by Philip Ross and others, applied mycelium to create leather-like materials for fashion, and by the late 2010s companies like MOGU were producing mycelium acoustic panels and flooring for interiors (bringing fungal materials into consumer product design). Government and industry also showed interest in fungi for sustainability—for example, NASA-supported concepts have proposed mycelium-based building components for space habitats. By the 2020s, mycelium composites evolved from niche experiments to a vibrant field: international research consortia (e.g., the FUNGAR project in the EU) investigate fungal architecture, and multiple companies mass-produce mycelium materials for packaging, textiles, and construction.

The historical trajectory above shows a shift from foundational inventions to broader adoption and scale-up. Early pioneers demonstrated that mycelium can "grow" novel materials, and over roughly 15 years, the field matured into a multidisciplinary endeavor spanning architecture, product design, and biotechnology. Today, mycelium composites are recognized for their potential to transform material economies toward circularity by literally growing replacements for plastics, foams, leathers, and even building materials.

2.2 Manufacturing Process and Material Properties

Fabrication methods for mycelium composites, at least with regards to architectural applications, have diversified from simple casting in molds to advanced additive manufacturing. Initially, the dominant technique was formwork casting: packing a mixture of fungal inoculum and plant-based substrate into a mold, incubating it until the mycelium binds the substrate into a solid shape, then drying the piece. This approach yields blocks, bricks, or panels in the mold's shape and was used in early projects like Hy-Fi and MycoTree, which grew mycelium in brick-shaped molds. Mold-grown composites are relatively easy to produce and were crucial for proving structural applications; however, they are limited to the geometric scope of the formwork and often have uniform properties. The internal structure (and thus performance) of such pieces can be hard to control, as nutrient and oxygen distribution are constrained by the mold geometry.

To overcome these limitations, researchers have turned to additive manufacturing (3D printing) with mycelium. Additive methods extrude a paste-like "mycelium ink" through a nozzle, depositing it layer by layer to build complex forms without a full mold. This allows the creation of intricate geometries, internal lattices, and graded structures that were previously unattainable. For instance, large-scale robotic extrusion of mycelium material was demonstrated by Goidea et al. [4] and Elsacker et al. [5], who systematically studied how paste composition, nozzle design, and print parameters affect fungal viability during printing. Their work established a methodology for printing sizable mycelium components (using a 4- or 6-axis robot) while keeping the fungus alive throughout the process. Another study by Gomaa et al. [6] investigated parametric 3D-printed geometries (using *Ganoderma* mycelium with hemp/straw substrate) and showed that print pattern and infill geometry significantly influence the composite's growth and strength—optimized lattice structures promoted thicker mycelial networks and improved mechanical performance. These advances suggest that geometry can be used as a tool to tune biological and structural outcomes in mycelium composites, an idea further extended by exploratory concepts like Lipińska et al. [7] who envisioned 3D printing mycelium with Martian soil for extraterrestrial habitats.

In parallel, hybrid fabrication techniques have emerged, combining mycelium with other materials to enhance properties. Researchers have embedded conventional reinforcement phases within growing mycelium matrices—for example, Wood-Veneer Reinforced Mycelium involves layering thin wood veneers inside a mycelial substrate before growth, yielding a plywood-like composite with far greater bending and tensile strength than mycelium alone. Özdemir et al. [8] demonstrated that such hybrids can meet requirements for architectural elements by leveraging mycelium as a natural adhesive binding the veneers. Another hybrid strategy is textile formwork: Kaiser et al. [9] developed a mycelium-infused paste that is cast onto knitted textiles, effectively "bio-plastering" a pre-shaped fabric form. The mycelium infiltrates and bonds with the textile, and once grown and dried, the composite (termed BioKnit) has a reinforced, tubular structure with improved load capacity (the team built a 1.8 m arched dome to prove its strength). Similarly, MYCOlullose, a 2024 experiment by Piórecka et al. [10], combined mycelium with bacterial cellulose (SCOBY) to create a multi-organism material, where a robust cellulose membrane complements the mycelial foam, resulting in a biohybrid with unique mechanical and hydration properties. Table 1 summarizes the major manufacturing approaches for mycelium composites and representative examples.

After fabrication, material properties of mycelium composites are evaluated in terms of mechanical performance, thermal behavior, and biological/aging characteristics. Because these materials are novel and biologically derived, their properties can vary widely depending on fungal species, substrate, density, and post-processing [11]. Nevertheless, a growing body of peer-reviewed literature provides insight into typical ranges:

Mechanical Properties: Mycelium composites are lightweight ($<$ 0.3–0.5 g/cm^3 density typically) and have moderate strength suitable for non-structural and some semi-structural uses. In compression, most pure mycelium materials exhibit strength on the order of a few hundred kPa [12]. These values are low compared

Table 1 Mycelium composite fabrication approaches and examples

Fabrication approach	Description and features	Example study
Formwork casting (static)	Grow mycelium in a fixed mold to obtain a preset shape. Suitable for blocks, bricks, panels; limited complexity	Hy-Fi mycelium bricks (The Living, 2014); MycoTree (KIT/ETH, 2017)
3D printing (extrusion)	Extrude a mycelium-rich paste layer by layer to build complex or large geometries without full molds. Allows internal channels (for aeration) and custom lattice infill	Pulp faction/protomycokion (bioDigital matter, Lund University, 2020); Robotic printing of mycelium (Vrije Universiteit Brussel 2022); m{AI}celium (University of Stuttgart, 2024)
Hybrid composite formation	Integrate mycelium with secondary materials (fibers, fabrics, etc.) during growth to reinforce the composite. Leverages mycelium as a biological binder	Wood-veneer-reinforced mycelium composites (University of Kassel, 2022); BioKnit (Newcastle University, 2023); Urban MYCOskin (UCL, 2024)

to conventional masonry or wood, but are comparable to foamed polymers. Importantly, mechanical properties can be improved by densification and additives: physically pressing mycelium boards can roughly double their strength (as the mycelium matrix is compacted), and as noted earlier, incorporating reinforcements (like veneers or textiles) can raise tensile and bending capacity significantly. For instance, the wood veneer–mycelium composites achieved substantially higher flexural strength than plain mycelium, making them viable as panels or beams. Mycelium materials are generally stiff enough in compression for load-sharing in lightweight construction, but their load capacity remains much lower than steel, concrete, or engineered timber, which is a known limitation.

Thermal and Fire Properties: A major advantage of mycelium composites is their thermal insulation performance. The porous mycelial structure traps air, yielding thermal conductivities typically in the range 0.04–0.13 W/m·K, comparable to traditional insulation materials like EPS (expanded polystyrene) or mineral wool. Additionally, mycelium-based boards tend to have inherent fire resistance: they can exhibit self-extinguishing behavior because they primarily char rather than sustain a flame. Tests have shown that mycelium composites ignite no faster than polystyrene and release far less heat and smoke when burned. In one comparison, mycelium materials produced much lower peak heat release than synthetic foam, indicating a potential safety advantage [13]. This fire performance is attributed to the biological composition—the high-carbon (char) content in mycelium causes it to form an insulating char layer when exposed to flame, rather than melting or dripping. Overall, the combination of insulative capacity and fire resistance makes mycelium composites very promising for building insulation and safing applications.

Biological and Environmental Properties: Being fungi-grown, these composites are biodegradable and compostable by design. At end-of-life, a mycelium composite can be crumbled and added to soil or compost where it will naturally decompose (typically within months to a couple of years, depending on thickness and conditions). This end-of-life scenario is a stark contrast to petrochemical foams or plastics, which often persist for decades or centuries. The mycelium itself and the agricultural fibers it binds are both organic matter that can be returned to the earth with minimal pollution, aligning with cradle-to-cradle sustainability principles. Moreover, the production of mycelium materials generally has a low carbon footprint: growth occurs at room temperature with fungi metabolizing waste biomass, consuming far less energy than the high-heat processes for synthetic materials. Life-cycle assessments have found mycelium insulation can outperform polystyrene, foam concrete, or mineral wool in terms of climate change impact, given its renewable feedstock and carbon sequestration during growth [11]. However, there are also biological challenges: mycelium composites are sensitive to moisture in service—if they get wet and are not properly sealed, they can be susceptible to rot or infestation by other molds (since they are essentially organic matter). To mitigate this, products often undergo drying and inactivation (heat-killing the fungus) once the desired form is grown. A fully dried mycelium composite is dimensionally stable and will not sporulate or grow further, behaving like a wood product. Still, moisture protection (via coatings or hydrophobic treatments) is usually recommended to ensure longevity, much as one would protect natural wood. Another consideration is variability—small changes in growth conditions can lead to differences in density or strength, so quality control in manufacturing is crucial. Ongoing research is addressing these issues, for instance by developing treatments to reduce water absorption, by genetically selecting fungal strains for more consistent performance, or by integrating metamaterial properties that reduce susceptibility to moisture through designed geometries. Adaptive design of the internal structures can simultaneously deter water ingress and facilitate drying. (Such functionally graded materials are a significant motivation and driver for additive fabrication and computational design in these contexts, and are discussed below as the "transcalar design process.")

2.3 *Applications beyond Architecture*

The use of mycelium-based materials has rapidly expanded in other industrial fields, finding applications in packaging, product design, and food. These developments not only demonstrate the versatility of fungal composites, but also reinforce public acceptance, supply chains, and interdisciplinary innovations that support their use in construction.

Sustainable packaging has become one of the earliest and most commercially successful applications of mycelium composites. Companies such as Ecovative pioneered this domain by growing fungal mycelium into custom-shaped packaging inserts using agricultural waste substrates. These foam-like materials provide comparable protection to expanded polystyrene but fully biodegrade after use, often within a matter of weeks under composting conditions. Global brands including Dell and IKEA have incorporated or attempted to incorporate this technology into their logistics chains, citing reductions in plastic waste and improved life cycle performance. (Since the announcement, IKEA has stated they "found it too difficult for manufacturers to produce the scale of packaging the company needs," and are focusing on other avenues to reduce plastic packaging for the time being [14].) In addition to electronics and furniture, these materials are increasingly used in cold-chain packaging and wine shipping, where insulation and impact resistance are critical. Their success in this domain has helped validate mycelium composites as reliable, industrial-grade materials.

Consumer products and interior design also make use of mycelium's unique tactile and aesthetic qualities. Designers have developed lamps, partition walls, and home furnishings that showcase the organic, fibrous textures of fungal materials. Companies such as MOGU have advanced this approach by offering commercially available acoustic panels and interior surfaces made from mycelium grown on textile waste. These products combine sound-dampening properties with fire resistance and compostability, aligning well with circular design principles.

A particularly dynamic field is mycelium-based textiles and leather alternatives. Startups like MycoWorks and Bolt Threads have developed fungal materials that mimic the look, feel, and mechanical performance of animal leather. MycoWorks' "Reishi" is composed of densely interwoven mycelial fibers that can be processed and dyed like traditional hide, and has been used in high-end handbags and accessories. These materials are grown in sheets under controlled conditions and require significantly less water and chemical treatment than conventional leather. They are now being integrated into fashion collections by luxury brands, offering a compelling combination of sustainability, performance, and design innovation.

Food and nutrition represent another major domain where fungal mycelium has found commercial success. The most notable example is Quorn, a mycoprotein product derived from *Fusarium venenatum*. Developed in the UK during the 1980s, Quorn was one of the first large-scale uses of fungal biomass as a direct food source. Grown in large aerobic bioreactors, the resulting mycelium is harvested, heat-treated, and texturized to mimic the structure and mouthfeel of meat. The product has been widely adopted as a sustainable protein alternative in vegetarian and vegan diets, particularly in Europe and North America. More recently, companies like Meati and MyForest Food have extended this concept to create whole-cut mycelium "steaks" and bacon substitutes, produced using similar fermentation processes. These applications exemplify the nutritional potential of mycelium and illustrate its capacity to form structured, protein-rich materials at scale.

Taken together, these developments position mycelium composites as a truly cross-sector innovation. In each case—packaging, interior products, fashion, or food—the material's biodegradability, moldability, and minimal resource demands are central to its value. Importantly, these non-architectural uses also feed back into the architectural field: they help build public familiarity, reduce production costs, and stimulate new processing techniques. The convergence of ecological benefits, design opportunities, and technological scalability makes mycelium an increasingly attractive material not just for buildings, but for the broader material culture of a circular economy.

2.4 Transcalar Complexity in Biodesign

Bringing biology into architecture involves bridging vast differences in scale—from enzymes rearranging cellulose chains at the nano-level to the meter- or even city-scale of a completed building. Architecture, especially when working with living matter, needs to embrace a "woven-together" perspective where interactions and feedback loops traverse multiple orders of magnitude.

Designing at one scale without acknowledging the others quickly runs into conflicts. For instance, if mycelial growth is suppressed at the micro-scale due to insufficient moisture, the macro-scale structure could fail to coalesce or meet performance targets. Conversely, an excessive focus on uniform engineering properties might ignore emergent local differentiation that could yield enhanced structural or functional outcomes. This transcalar interdependence favors the use of computational design workflows where geometry, material behavior, and biological processes interact dynamically.

The synergy of biology, computation, and fabrication strategies forms the basis for novel forms of material computation, in which the internal complexities of living substrates are not abstracted away but integrated into the design logic. This perspective is crucial to the approach used in Protomycokion, discussed below, where digital tools orchestrate the shape of a fungal-lignocellulosic composite so that the fungus can properly colonize the material while still meeting architectural aspirations.

3 The Transcalar Design Approach

A unifying challenge in mycelium-based architectural fabrication is ensuring that processes at very small scales, such as biochemical transformations, align with structural and environmental goals at the larger scale, and that the overall geometry facilitates conditions for growth at the small end. Depending on the specific approach the nature of this scalar interaction varies, but in the case of Protomycokion [15] it manifested as:

Molecular/nanoscale: Fungal enzymes break down lignin or cellulose, producing new polymeric substances like extracellular polysaccharides, chitin, or proteins. At this scale, minor changes in pH, moisture, and nutrient availability can drastically affect growth rates. These transformations may also enable, for instance through bio-welding, localized repairs or reconfigurations of a structure.

Micro and milli-scale: Mycelial hyphae grow and branch in three dimensions, binding together substrate particles. Features such as pore size or fiber orientation influence colonization and, consequently, material cohesion and strength.

Meso-scale: At the component level (e.g., a 3D-printed block, panel, or column segment), geometry significantly impacts airflow, hydration, and mechanical behavior. These effects propagate both upward and downward through the scales.

Macro-scale: Assemblies or entire buildings must integrate these biofabricated components without compromising durability, safety, or occupant comfort. For example, the channels and cavities that are necessitated for the fungal growth, may provide various means to control the building's internal climate.

These nested scales offer a conceptual map for understanding fungal-material interaction, but they also serve as a framework for actively shaping performance. By designing across these layers, materials can be tuned to respond to both biological needs, mechanical constraints, and architectural ambitions.

Through engagement with transcalar relationships, it becomes possible to fabricate structures that are functionally graded, that is, their behavior and performance can be locally adapted rather than uniform throughout. In fungal composites, this principle can be leveraged to regulate moisture, a critical factor for both viability during fabrication and long-term durability. Meso-scale geometries, for instance, can be designed to support passive airflow, allowing structures to dry efficiently after exposure while also shedding water to prevent ingress. Computational methods enable the fine-tuning of airflow direction and intensity, responding to the specific demands of different regions within a component. Combined with micro-scale features such as hygroscopicity or hydrophobic properties of the hyphae, and macro-scale considerations of orientation and exposure, this integrated approach can significantly enhance performance. These strategies rely on digital design workflows that coordinate biological growth and environmental behavior—topics explored further in the following section.

3.1 Computational Tools and Biodesign Methodologies

Algorithmic design environments are central to managing these complexities. Reaction-diffusion algorithms (as used in Protomycokion) or other generative models can specify topologies with high surface-to-volume ratios, enabling fungus to flourish. Multi-agent simulations can track how printing trajectories or humidity conditions shift over time, and parametric design systems allow for real-time updates when, for instance, the substrate's moisture content changes or the fungus is found to require different aeration channels. Additionally, various simulation techniques may help predict how partially colonized, water-rich materials will behave as they dry and become more rigid. This is potentially critical, as the locally sensitive design means that conventional performance metrics, based on uniformity and standardization, may not be sufficient.

Crucially, these tools need to operate beyond the typical linear engineering logic. Mycelium-based materials often exhibit nonlinearity: small changes in local temperature or substrate composition may cause large variations in final compressive strength or shrinkage rates. Because living mycelium is dynamic, designers must often treat the printed object not as a final artifact but as a transitional form that evolves before stabilizing. This evolution can be harnessed to "bio-weld" components together, or to create controlled shrinkage patterns that lock components into stable forms.

3.2 Collaborative, Cross-Disciplinary Processes

The success of these methods depends on close collaboration across disciplines. Microbiologists investigate strain selection and growth conditions; materials scientists address rheology, mechanical characterization, and hybridization with fibers or fabrics; architects create generative designs that orchestrate geometry and ecological performance.

Such a broad collaboration is not just a matter of logistical convenience but is essential to navigate the intricacies of scale, variability, and biological agency. Research-by-design projects like Protomycokion or BioKnit thus reflect "living" laboratories, where iterative experimentation with prototypes and demonstrations feeds back into computational models, ultimately informing the next round of design. Academic programs, architectural robotics labs, and specialized startups are increasingly pursuing this collaborative model, leading to innovations not only in material science but also in the social and logistical aspects of how buildings are designed, manufactured, and maintained.

4 The Protomycokion Project

In exploring how fungal biotransformation and additive fabrication intersect, the Protomycokion column offers a valuable narrative [4, 15] Here, we see how the interplay of living matters and digital design gives rise to an architectural prototype that articulates the transcalar approach.

4.1 Materials and Methods

The substrate for Protomycokion comprised a lignocellulosic mixture of wood chips, paper pulp, clay, and water, complemented by natural thickeners. This "pulp" was then inoculated with fungal strains selected for strong colonization and cohesive bonding (Fig. 1a–b). Once the mycelium was established, the substrate (or "live

Fig. 1 **a** Fungal mycelium propagation on agar plates, the preliminary cultivation of the fungus. **b** Inoculation of the lignocellulosic pulp with the blended mycelium culture. **c** Layered deposition modeling (LDM) of the inoculated pulp, with the paste extruded through a customized nozzle. **d** Incubation of the printed component in a controlled, high-humidity environment, allowing the fungus to colonize and bond the substrate. (Reproduced from [15] under CC-BY 4.0)

pulp") became extrudable through a 3D printing nozzle, depositing thin layers that conformed to precomputed toolpaths (Fig. 1c).

This process inquired into numerous unknowns: how wide should the nozzle be to balance deposition speed with adequate oxygenation? How moist could the pulp remain to ensure fungal viability without collapsing the print layers? How should layers interlock to avoid delamination? How can toolpaths be organized and shaped to maximize stability of the wet pulp?

Protomycokion's geometry emerged from parametric scripts that balance structural logic, fungal physiology, and the aesthetic or spatial aims of the column. Reaction-diffusion algorithms generated internal porosity and high surface area, while subsequent algorithms adapted the geometries to the requirements of the printing process (Fig. 2). Meanwhile, the approach to printing was carefully phased: first, the "live pulp" was extruded; then, the piece incubated in a high-humidity environment to allow the fungus to bind and partially transform the substrate (Fig. 1c–d); finally, the column was gradually dried, halting further growth and stabilizing the composite, and assembled. The process was set up to enable bio-welding in the assembly, effectively creating a 2-stage layer-by-layer print process operating self-similarly at two nested scales. The ability of living cells to effect molecular reassembly means that such a process can potentially overcome the challenges

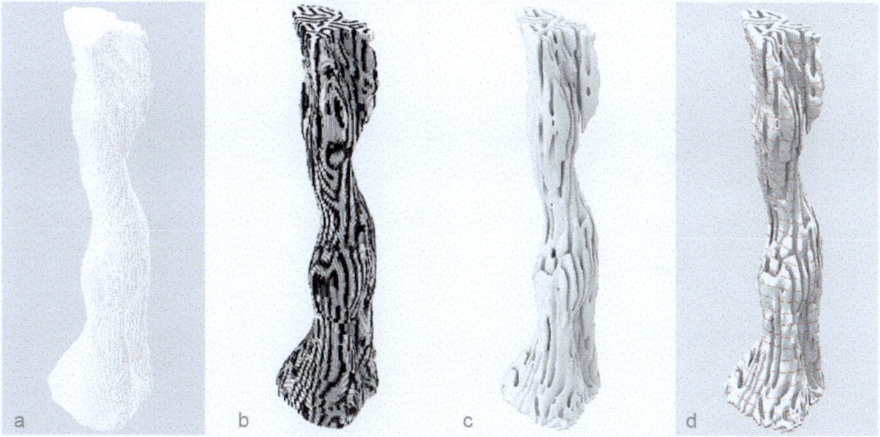

Fig. 2 Algorithmic design of the Protomycokion column. A reaction-diffusion model **b** provides semi-random, high-surface-area geometries within a defined boundary volume **a**. The resulting voxel space is then translated into parametric tool paths suitable for LDM extrusion **c-d**. This process balances structural needs, fungal growth requirements, and aesthetic goals. (Adapted from [4] under CC-BY 4.0)

of bond strength between previously cured layers commonly experienced in, e.g., 3D-printed concrete.

4.2 Outcomes and Observations

The final column revealed a distinctive woven-like surface where fungal hyphae had "knit" layers into an interlocking tapestry (Fig. 3). While the individual segments of the column were quite varied in their appearance and some mismatch occurred, the overall perception of the column is one of consistency and variability; many have noted the apparent similarities to biological structures such as trees. We would argue that these similarities emerge not primarily from the use of wooden raw materials, but rather from the spatial continuity of emergent, process-based design across the scalar-spatial spectrum.

Mechanical testing showed that while structural strength was promising, these properties lagged more refined composites and conventional construction materials. Still, the synergy of living growth and computational geometry made possible the bridging of micro-level viability and macro-level stability. This resonates with findings from other research. For example, the BioKnit team improved tensile properties by pairing fungus with a permanent knitted textile; likewise, Wood-Veneer-Reinforced Mycelium Composites managed better bending resistance via veneer layers.

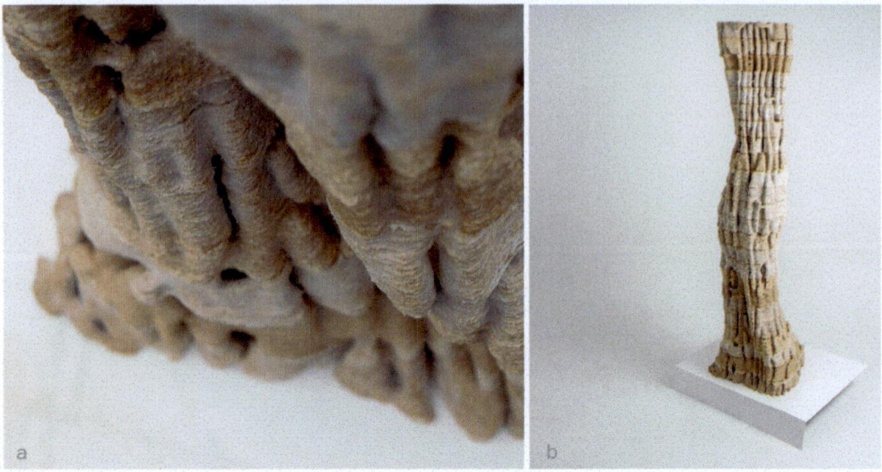

Fig. 3 **a** Close-up view of the column texture and surface, formed by the intersecting printed layers and hyphal growth. **b** The completed Protomycokion column after printing, colonization, and assembly. (Reproduced under CC-BY 4.0)

Protomycokion, however, aimed less at maximizing any single property and more at demonstrating a holistic, transcalar design-fabrication pipeline. In so doing, it underscored both the potential and the shortcomings: the fungus's dynamic behavior can lead to unpredictable shrinkage, local distortions, or non-uniform mechanical strength. Future refinements might involve controlling humidity with greater precision or partially reinforcing the pulp with secondary scaffolds, echoing the multi-material approach seen in other recent projects.

5 A Transformation of Architectural Design

The experiences gleaned from Protomycokion and its predecessors suggest broader principles for integrating fungal biofabrication into architectural practice [16]. While these themes often appear in separate frameworks (e.g., "material ecology," "biomimetics," "computational design"), they converge in the emergent reality of living, 3D-printed architecture.

5.1 Diversity in Material Sourcing and Design

Mycelium can thrive on a variety of lignocellulosic substrates (agricultural residues, forestry by-products, domestic and construction waste, invasive plant biomass, etc.) and there exists a near limitless variety of fungal strains. The capacity for fungi to convert a wide array of lignocellulosic feedstocks into stable construction materials is a remarkable advantage. Instead of relying on uniform, industrially processed inputs (e.g., purely wood fiber from a single source), local or regional waste streams can be harnessed, thereby reducing transportation emissions and supporting decentralized manufacturing.

Moreover, a mixture of fungal species may provide specialized advantages. Some fungi excel at hydrophobic coatings, while others produce denser networks of chitin that enhance mechanical performance. In principle, an ecosystem of multiple microbial agents could lead to sophisticated composites with gradients of properties. Such possibilities amplify the design toolkit but also demand advanced process control.

Architects and builders thus gain an opportunity to connect intimately with local ecological contexts. For example, in agricultural regions with abundant straw or rice husks, mycelium-based panels and blocks grown from such waste could integrate seamlessly into local building cultures. In other parts of the world, invasive plant species might be transformed from an ecological burden into valuable building material, all through targeted fungal processes.

5.2 *Complexity and Specificity of Form*

Biological forms are often highly intricate, reflecting the complexity of their internal organization and the specific functions they serve. When architects incorporate living organisms, complexity ceases to be a mere aesthetic choice: it becomes a functional imperative. For instance, in mycelium-based designs, thin-walled curvilinear geometries help supply oxygen to the fungus and reduce the risk of internal rotting or incomplete colonization. Complex folded structures can also distribute stress more evenly and increase insulation performance.

Such specificity of form is increasingly enabled by additive manufacturing, which can realize shapes that defy the constraints of traditional formwork or molds. Projects like MycoTree and BioKnit illustrate the potential to craft freeform structural components, while smaller-scale design explorations have shown how generating fractal or cellular geometries can optimize strength-to-weight ratios in mycelium-based composites.

Moving away from the standardized, rectangular block approach can have profound implications. Buildings can be functionally graded, with certain regions denser than others, or with fluid channels integrated for microclimate regulation. This echoes the approach seen in vernacular or traditional earthen architectures, where local availability and building know-how once drove the creation of complex, site-responsive forms.

5.3 Durability through Resilience

The notion of "durability" in modern architecture typically centers on materials that resist change: steel beams that do not corrode, concrete that remains inert under compression, and so forth. By contrast, living materials or those produced through living processes often challenge these assumptions by embracing resilience—the capacity to adapt, regenerate, or be repaired.

Mycelium composites, while robust in many respects, might degrade in certain climates if left untreated. Designing for resilience includes acknowledging that parts of a building could be periodically regrown or replaced. Instead of sealing the material away under synthetic coatings, we advocate, at least in certain contexts and scenarios, for maintainable envelopes and interior components that can be disassembled, composted, and reprinted or regrown. This cyclical approach mimics natural ecosystems, where constant renewal does not imply inferior durability, but rather cyclical, adaptive longevity.

In addition, structural resilience can be fostered by geometry or redundant load-bearing paths. A 3D-printed lattice of mycelium composite can be designed to maintain integrity even if certain segments fail or degrade. This moves architecture away from single-point, catastrophic failure modes and more toward the layered resilience we see in biological systems.

5.4 Feedback and Adaptation

Biological organisms thrive on feedback loops, continually adapting to environmental pressures. Mycelium-based building components, in principle, can emulate these loops at several levels.

On a material level, if a component is damaged or exposed to changing conditions, fungal tissues might seal cracks or produce additional protective coatings, in much the same way coral reefs respond to stress by growing more robustly. While buildings are unlikely to be constructed with fully active, living fungus, dormant cells could potentially be embedded, enabling them to re-activate under certain stimuli, like the self-healing strategies now being explored in concrete technology [17]. From a design perspective, sensors for temperature, humidity, and mechanical stress may be installed to provide data that feed back into computational models, thereby guiding subsequent design iterations or even in-situ robotic re-printing. Finally, at the scale of the building's overall evolution, the inherent capacity of fungal organisms to transform and reorganize material offers new possibilities for continuous change over time. Bio-welding, for instance, might enable seamless incorporation of new or modified components, while damaged or unwanted parts could be recycled into fresh building material.

Although truly self-repairing buildings remain a future prospect, smaller prototypes in controlled environments have already demonstrated partial feedback loops.

Over time, such innovations could yield buildings that not only endure but also adapt to seasonal shifts, occupant demands, and the slow accumulation of wear and tear.

6 Rethinking Construction Practice

Mycelium-based 3D printing signals a departure from high-carbon or energy-intensive building materials like concrete, steel, and petro-plastics. Yet this approach also requires rethinking nearly every facet of the construction process:

Time and Construction Sequence: Traditional construction often aims for speed; by contrast, mycelium composites need days or weeks to colonize, plus time for drying. This shift challenges project timelines but could be mitigated by staging biofabrication early. Furthermore, such a perspective shift in the construction timeline may well enable a local, ongoing production of the raw materials—perhaps even on or within the building itself.

Material Production: Rather than mass factories producing identical standardized products, we imagine networks of smaller facilities or even on-site labs cultivating fungi in local substrates. Such approaches also enable sustainable implementation of continuous adaptation and renewal of the building to fit evolving needs.

Maintenance and Lifecycle: Fungal-lignocellulosic structures might degrade or lose performance if not carefully maintained. However, the same biological processes that might degrade them can also recycle them into new components. Further, biomaterials do not necessarily exclude other construction materials, wither conventional or vernacular—materials such as stone, concrete, earth, or glass may complement biomaterials, particularly where these materials lack.

This shift is not only environmental but cultural. Communities, clients, and end-users must adjust to the idea that buildings could be partially alive, or at least shaped by living processes. Some might enthusiastically embrace the story of hyper-local, biodegradable walls; others might worry about mold, spores, or perceived impermanence. Education, transparent testing, and regulatory frameworks will all play major roles in acceptance.

6.1 Challenges in Implementing Biodesign

Implementing biodesign at scale brings forth numerous technical and material challenges. Foremost among these is controlling growth: ensuring consistent results in large-scale production proves difficult when dealing with living organisms. Even small variations in humidity or substrate composition can trigger significant swings in outcomes. Another key issue relates to structural certification, since building codes and engineering standards often do not address biologically derived materials and unique, one-off components.

Regulatory and cultural barriers also come into play. Engineers and building officials usually demand performance data in standardized formats, and mycelium-based products must therefore undergo extensive testing to verify fire safety, load capacity, and durability. Public perception further complicates the adoption of living materials, as many people express concerns about hygiene, toxins, or decay—despite evidence showing these challenges can be effectively managed.

A third area of difficulty involves scalability and commercialization. Although the inherent variability of fungal growth can be viewed as a strength, most industrial processes require a degree of uniformity and predictability to remain cost-effective. In addition, establishing reliable supply chains for regionally sourced biomass and fungal strains often necessitates forging partnerships among agriculture, waste management, and construction sectors, each of which operates with distinct norms and logistical constraints.

There are ethical dimensions to growing living matters in mass quantities, akin to broader debates over biotech and industrial-scale agriculture. The transcalar lens highlights how large-scale expansions of fungal-based industries could disrupt local ecologies if not managed carefully. However, integrated supply chains—where substrate is an agricultural by-product, and post-use composting re-enriches soils—can make these processes far less extractive than conventional building materials. The synergy of local nutrient loops, minimal external energy input, and potential carbon sequestration offers a strong environmental rationale for further development.

6.2 Future Directions

As more projects merge fungi with other organisms (like algae or bacteria), new synergies may arise. Multifunctional "biohybrid" systems could incorporate advanced digital sensing for real-time adaptation or harness the emergent patterns of different species for specialized tasks. Mycelium-based printing might also move beyond single-phase processes toward multi-phase approaches, layering different inks or employing partially cured biomaterials in a continuous cycle, and alternatives to LDM techniques, such as particle-bed printing, can enable greater geometric freedom.

The number of fungal species that have currently been explored in this context relative to the number of existing potential species and strains is so far very small. The properties that may be achieved through alternative strains can vary greatly, as is illustrated by the material properties of Protomycokion which was fabricated using a brown-rot fungus rather than the typical white-rot: its surface hardness as well as its appearance and aesthetic are profoundly different from many other fungus-based materials (Fig. 4). Even beyond this, various forms of genetic engineering may enable us to tune (or radically alter) the behavior and output of the fungus we use.

Fig. 4 Early prototypes fabricated from three different pulp conditions: **a** un-inoculated pulp, **b** pulp inoculated with a white-rot fungus, and **c** pulp inoculated with a brown-rot fungus (*Gloeophyllum*). The *Gloeophyllum* strain produced a distinctly harder, more hydrophobic surface, suggesting possible advantages for outdoor or high-moisture architectural applications. (Reproduced from [4] under CC-BY 4.0)

This transcalar design space, which weaves together hundreds or thousands of individual parameters within a multitude of distinct process components, may be particularly well suited for, or even necessitate, the use of various artificial intelligence (AI) approaches. A traditional reductionist approach of isolating one variable at a time may not be feasible, or at least not to combine the millions of parameters in an optimal or functional way for a particular design task. While a (research-by-) design approach has proven successful at the initial task of untangling and demonstrating the potential of biodesign, the use of AI is likely necessary to implement such a paradigm in the construction sector, and efforts are already underway to bring such technologies into architectural design, either though machine learning [18], or agent-based, self-organizing modeling [19].

Eventually, the building site may resemble a controlled ecosystem rather than a purely mechanical assembly zone. Habitat design for space or other extreme environments can further push these ideas, as distance from Earth drives up the value of in-situ resource utilization and low-energy processes. Indeed, if we can build with fungal-lignocellulosic composites on a barren planet, bridging microgravity and minimal external resources, what might we achieve on Earth with a far richer variety of materials and environmental conditions?

7 Conclusion

Fungal biotransformation with additive fabrication signals a transformative chapter in architectural evolution. By coupling the innate capabilities of fungi—decomposing and recombining organic matter—with computational design and robotic control, architects and engineers gain an unprecedented ability to grow custom, biodegradable materials that span from small interior panels to entire buildings. Through a transcalar approach, we appreciate that design decisions at the micro-scale (enzyme production, mycelial networking) reverberate up to the macro-scale (structural integrity, occupant comfort), and vice versa.

Over the last decade, pioneering projects such as Hy-Fi, Protomycokion, and BioKnit have steadily demonstrated how living composites can be formed, manipulated, and integrated into real-world contexts. Early difficulties in mechanical reliability and uniformity have sparked new strategies, from hybrid reinforcement (wood veneers, fabrics) to dynamic printing processes. Meanwhile, parallel work has introduced design concepts for entire habitats—on Earth or off-planet—grown from mycelium substrates, imagining a near future where "construction" is partially replaced by "cultivation." Yet bringing these visions to fruition demands an equally fundamental rethinking of the design process itself.

Looking forward, it becomes clear that adopting biological paradigms will require moving beyond conventional linear frameworks in which a building is meticulously designed, erected, used, and ultimately demolished. In their place, we can envision more fluid, biology-like methods, where the boundary between designing, occupying, and continually renewing architecture is blurred. In this model, buildings would be seen as living systems, undergoing periodic regrowth or adaptation rather than awaiting demolition. Construction sites might become local ecologies of fabrication, where materials are cultivated in response to shifting demands, and the building itself evolves alongside its inhabitants. Such an ongoing, iterative process of making stands to foster more sustainable, resilient, and context-attuned environments, redefining how architecture is conceived, inhabited, and perpetually reimagined.

As these prototypes and experiments accumulate, we begin to glimpse a fundamental shift in how architects and communities conceive of building. Rather than inert, carbon-heavy materials, we see ephemeral, circular cycles of growth and decomposition. Rather than the forced uniformity of mass production, we see emergent patterns shaped by local feedback loops. And rather than casting living organisms as subjects for imitation, we invite them as partners in a genuinely collaborative form of making. The underlying impetus remains the same: to address an urgent environmental crisis while still providing functional, creative, and dignified spaces for humans. Mycelium-based biofabrication, with all its complexities and unknowns, suggests that the path to ecological harmony might well be paved—indeed, grown—by the humble yet extraordinarily capable kingdom of fungi.

References

1. Grover, A.: The future is fungal: Interview with Phil Ross
2. Nagy, D., Locke, J., Benjamin, D.: Computational brick stacking for constructing free-form structures. In: Modelling Behaviour, pp. 203–212 (2015). https://doi.org/10.1007/978-3-319-24208-8_17
3. Heisel, F., Lee, J., Schlesier, K., Rippmann, M., Saeidi, N., Javadian, A., Nugroho, A.R., Van Mele, T., Block, P., Hebel, D.E.: Design, cultivation and application of load-bearing mycelium components. Int. J. Sustain. Energy Dev. **6**, 296–303 (2017)
4. Goidea, A., Floudas, D., Andréen, D.: Pulp faction: 3D printed material assemblies through microbial biotransformation. In: Skavara, M., Burry, J., Sabin, J.E., Sheil, B. (eds.) Fabricate 2020: Making Resilient Architecture, pp. 42–49. UCL Press, London (2020)
5. Elsacker, E., Peeters, E., De Laet, L.: Large-scale robotic extrusion-based additive manufacturing with living mycelium materials. Sustain. Futures. **4**, 100085 (2022). https://doi.org/10.1016/j.sftr.2022.100085
6. Gomaa, H., Chau, W.M., Karazi, Y., Biala, E., Akbar, Z., Wortmann, T., Ostermann, M.: Influence of geometry on growth and strength of 3D-printed mycelium composites: a data-driven study. In: Scalable Disruptors, pp. 331–345 (2024). https://doi.org/10.1007/978-3-031-68275-9_27
7. Lipińska, M.B., Dade-Robertson, M., Zhang, M., Rothschild, L.J., Dade-Robertson, M.: Drop the base: biological, ISRU-based aleatory construction system for Martian habitats. In: 2023 International Conference on Environmental Systems (2023)
8. Özdemir, E., Saeidi, N., Javadian, A., Rossi, A., Nolte, N., Ren, S., Dwan, A., Acosta, I., Hebel, D.E., Wurm, J., Eversmann, P.: Wood-veneer-reinforced mycelium composites for sustainable building components. Biomimetics **7**, 39 (2022). https://doi.org/10.3390/biomimetics7020039
9. Kaiser, R., Bridgens, B., Elsacker, E., Scott, J.: BioKnit: development of mycelium paste for use with permanent textile formwork. Front. Bioeng. Biotechnol. **11**, 1229693 (2023). https://doi.org/10.3389/fbioe.2023.1229693
10. Piórecka, N.B., Scully, P., Salmane, A.K., Parker, B., Cruz, M.: MYCO-lullose: fabricating biohybrid material system with mycelium-based composites and bacterial cellulose. In: Sustainable Development Goals Series, Part. **F2790**, 597–614 (2024). https://doi.org/10.1007/978-3-031-36554-6_38
11. Camilleri, E., Narayan, S., Lingam, D., Blundell, R.: Mycelium-based composites: an updated comprehensive overview. Biotechnol. Adv. **79**, 108517 (2025). https://doi.org/10.1016/j.biotechadv.2025.108517
12. Alaneme, K.K., Anaele, J.U., Oke, T.M., Kareem, S.A., Adediran, M., Ajibuwa, O.A., Anabaranze, Y.O.: Mycelium-based composites: a review of their bio-fabrication procedures, material properties and potential for green building and construction applications. Alex. Eng. J. **83**, 234–250 (2023). https://doi.org/10.1016/j.aej.2023.10.012
13. Jones, M., Bhat, T., Huynh, T., Kandare, E., Yuen, R., Wang, C.H., John, S.: Waste-derived low-cost mycelium composite construction materials with improved fire safety. Fire Mater. **42**, 816–825 (2018). https://doi.org/10.1002/fam.2637
14. Lai, O.: Ikea to Eliminate Plastic from Consumer Packaging by 2028. https://earth.org/ikea-to-eliminate-plastic-from-consumer-packaging-by-2028/
15. Goidea, A., Floudas, D., Andréen, D.: Transcalar design: an approach to biodesign in the built environment. Infrastructures **7**, 50 (2022). https://doi.org/10.3390/infrastructures7040050
16. Andréen, D., Goidea, A.: Principles of biological design as a model for biodesign and biofabrication in architecture. Archit. Struct. Constr. **1**, 1–11 (2022). https://doi.org/10.1007/s44150-022-00049-6
17. de Rooij, M., Van Tittelboom, K., De Belie, N., Schlangen, E.: Self-Healing Phenomena in Cement-Based Materials. Springer, Netherlands, Dordrecht (2013)
18. Rossi, G., Chiujdea, R.S., Hochegger, L., Lharchi, A., Harding, J., Nicholas, P., Tamke, M., Thomsen, M.R.: Statistically modelling the curing of cellulose-based 3D printed components:

methods for material dataset composition, augmentation and encoding. In: Design Modelling Symposium Berlin 2022: Towards Radical Regeneration, pp. 487–500 (2022). https://doi.org/10.1007/978-3-031-13249-0_39
19. Goidea, A., Popescu, M., Johansson, A.T., Andréen, D.: Algorithmic modeling of functionally graded metamaterials in 3D printed building envelopes. Archit. Intell. **3**, 1–15 (2024). https://doi.org/10.1007/s44223-024-00068-x

David Andréen is Associate Professor in Architecture and director of the bioDigital Matter research group at Lund University, Sweden. His work explores the intersection of computational design, digital fabrication, and biological materials, aiming to redefine architecture's role in ecological transformation. David holds a doctorate from the Bartlett School of Architecture, UCL, and leads research integrating robotics, additive manufacturing, and bio-inspired design. His projects, including the Protomycokion and Meristem Wall, have been exhibited internationally, notably at the Venice Biennale. Committed to interdisciplinary collaboration and innovation, David actively contributes to advancing sustainable, adaptable, and biologically integrated approaches to architectural design.

Chapter 7
Three-dimensional Printing of Lignin

Md Shajedul Hoque Thakur, Md Tofazzal Hossain, Mir Muhammad Fahad, and Muhammad M. Rahman

Abstract Lignin, one of the most abundant biopolymers on Earth, has emerged as a promising material for multifunctional applications utilizing 3D printing techniques. As a low-cost and underutilized by-product of the pulp and paper industry, lignin offers an attractive alternative to petroleum-derived polymers that are commonly used in 3D printing applications. This chapter discusses the integration of lignin into 3D printing technologies, providing a comprehensive overview of the processing, properties, and applications of lignin-based systems. The chapter begins with an introduction to the biodegradability and chemical versatility of lignin, and its extraction, purification, various chemical modifications and blends. Next, we review its compatibility with various 3D printing techniques, such as fused filament fabrication (FFF), stereolithography (SLA), and direct ink writing (DIW). The printability challenges of utilizing lignin as a standalone material with each of these techniques are highlighted. The various approaches to enhance printability through blends with other polymers and rheology modifiers are discussed. The potential of lignin-cellulose composites is also explored, highlighting their synergistic properties for 3D printing. Our works on 3D printing of wood from lignin and cellulose combination via DIW technique of a water-based ink and enhancing their mechanical properties with natural fiber reinforcements are described. Finally, the chapter concludes with applications and possible future directions. Overall, this chapter underscores the transformative role of 3D printing in valorizing the abundant biopolymer lignin.

M. S. H. Thakur · M. T. Hossain · M. M. Fahad · M. M. Rahman (✉)
Department of Mechanical and Aerospace Engineering, University of Houston, Houston, TX, USA
e-mail: maksud@uh.edu

M. S. H. Thakur
e-mail: mthakur@uh.edu

M. T. Hossain
e-mail: mhossain19@uh.edu

M. M. Fahad
e-mail: mfahad2@uh.edu

© The Author(s), under exclusive license to Springer Nature Singapore Pte Ltd. 2026
C. Klemmt and G. Beaucage (eds.), *Biodegradable 3D-Printing*, The Materials Research Society Series, https://doi.org/10.1007/978-981-95-4025-9_7

Keywords 3D printing · Additive manufacturing · Lignin · Sustainability · Biocomposites

1 Introduction

As the world seeks sustainable alternatives to petroleum-based polymers, lignin, a naturally abundant biopolymer, presents a promising candidate for a wide range of multifunctional applications [1]. Lignin plays a crucial structural role in plant cell walls, providing structural rigidity, nutrient transport, and resistance to microbial degradation. As a by-product of pulp and paper mills, the potential of lignin has historically been underutilized, due to its lack of processibility. To date, most of it burned for energy rather than valorized into high-value products [1–3]. However, with the growing emphasis on sustainable materials and circular economy principles, lignin is gaining attention as a renewable feedstock for advanced material applications. Its uniquely complex aromatic heteropolymer structure and its variation based on plant species and growth conditions enable its potential for various multifunctional applications.

3D printing, also known as additive manufacturing (AM), presents a promising avenue for processing lignin into structural materials with tailored properties [4]. Lignin is a readily available and low-cost resource with an estimated global production of 100 million dry tons per year. Its price, ranging from 200 to 500 USD per dry ton, is significantly lower than that of synthetic polymers such as polyethylene (1000 USD per dry ton) [5]. Given this economic advantage, lignin has the potential to replace petroleum-derived synthetic polymers currently used in AM, including acrylonitrile butadiene styrene (ABS), polyethylene terephthalate (PET), polycarbonate (PC), and polypropylene (PP) [6]. Furthermore, lignin exhibits antioxidant and antibacterial properties, as well as natural biodegradability, that could enhance the functionality of 3D-printed materials. These attributes make lignin an attractive candidate for sustainable 3D printing applications, reducing reliance on petroleum-based polymers while introducing multifunctional benefits.

Despite its potential, lignin's inherent processing limitations pose significant challenges for 3D printing applications. Unlike engineered polymers designed for manufacturing, lignin lacks the processability required for seamless integration into standard AM workflows. Lignin is a complex, heterogeneous polymer composed primarily of phenylpropanoid units linked through various ether and carbon–carbon bonds. The chemical composition and molecular weight distribution of lignin depend heavily on its extraction method, leading to significant variations in processability and material properties. The diversity of lignin sources, such as kraft lignin, organosolv lignin, soda lignin, and hydrolysis lignin, results in different thermal stability, molecular weight, and solubility characteristics, all of which influence its suitability for various3D printing techniques. Researchers have explored several strategies, such as polymer blending, chemical modification, and composite formulations, to address these challenges and enhance lignin's compatibility with 3D printing technologies [6–8].

This chapter explores such diverse approaches to integrating lignin into 3D printing technologies. It begins by discussing the chemical structure and fundamental properties of lignin its compatibility with different 3D printing techniques, including fused filament fabrication (FFF), stereolithography (SLA), and direct ink writing (DIW). Next, the challenges and solutions associated with processing lignin for these AM techniques are discussed. The final section of this chapter highlights the advancements in 3D printing of wood-like materials from lignin, showcasing the potential of lignin-based 3D printing in the development of high-performance bio-based materials.

2 Lignin Processing

In 1838, Anselme Payen, a French chemist, treated wood with nitric acid and sodium hydroxide and obtained two separate substances: cellulose and a high carbon solution in which cellulose was soaked, called "encrusting material." Later, in 1865, Schulze named this material "lignin" in 1865, from the Latin word "*Lignum,*" which means wood. In 1890, Benedikt and Bamberger discovered that lignin was primarily found in the cell wall of lignocellulosic materials, such as wood, and composed of more functional groups, i.e., methoxyl lignin, which also varies from source to source (Fig. 1a) [9].

Naturally, lignin is formed through a lignification process in plants, where the three units of monolignol (4-hydroxyphenylpropanoids)precursors go through an oxidative polymerization reaction. These monolignols are (a) p-coumaryl, (b) coniferyl, and (c) sinapyl alcohols. These monolignols differ according to their degree of methoxylation at the C3 and C5 of the aromatic ring and create the three units of lignin polymer chain: p-hydroxyphenyl (H), guaiacyl (G), and syringyl (S), connected through mostly β-O-4 ether linkage along with other covalent bonds, such as 4-O-5 biphenyl ether, β-β resinol, and β-5 phenyl coumaran. This polymerization process forms a highly branched, three-dimensional, heterogeneous polymer structure (Fig. 1b) [11].

2.1 Extraction and Purification of Lignin

The primary source of lignin is cellulosic biomass (such as softwood, hardwood, and grass), which is produced as a by-product of the various extraction and pretreatment processes for pulp or second-generation ethanol production. Based on the isolation techniques of lignin, its functional group can vary, which will alter its chemical composition, molecular structure, and molecular weight. As a result, physical properties, such as solubility, viscosity, wettability, etc., will also be changed. Hence,

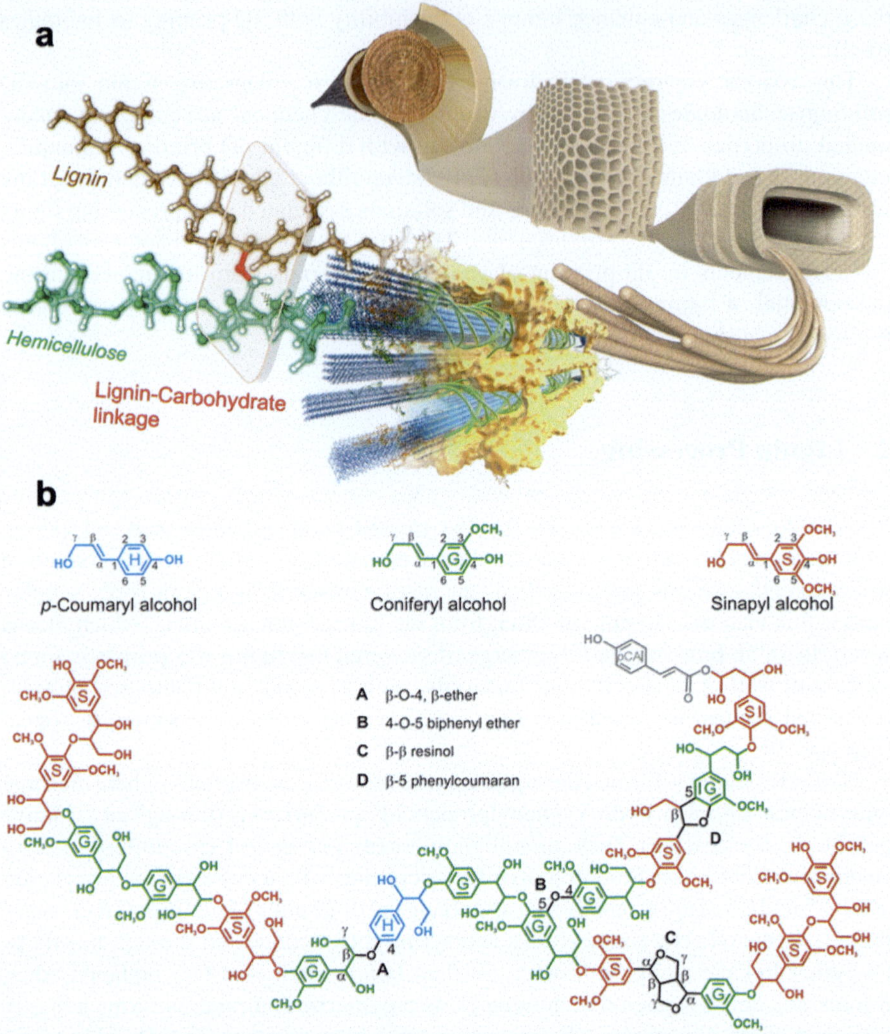

Fig. 1 **a** Depiction of the co-existence of cellulose, hemicellulose, and lignin in the cell wall of a lignocellulosic substrate. Reproduced under the terms of the creative commons attribution license [10]. **b** Chemical structure of lignin's original three constituent monolignols: p-coumaryl alcohol (H unit), coniferyl alcohol (G unit), sinapyl alcohol (S unit), and the final lignin polymer. Reproduced under the terms of the creative commons attribution license [11]

the lignin extraction procedure differs for each specific application [12]. Regardless of the extraction process, the lignin polymer chain gradually breaks down into shorter molecular structures by breaking the aliphatic -OH groups, β-O-4, and β-β bonds. Based on the isolation process, lignin can be broadly classified into two types: sulfur-containing lignin and sulfur-free lignin [13].

2.1.1 Sulfur-Containing Lignin

Kraft Lignin

The kraft process is essentially a high-temperature, high-alkaline medium delignification process, where the lignocellulosic raw materials are boiled with sodium hydroxide (NaOH) and sodium sulfide (Na_2S) at 140–170 °C. This process produces a black liquor consisting of degraded lignin and some other organic and oxidized inorganic compounds. The lignin is isolated from black liquor through an acidification process, which results in an abundance of C-C bonds and ether bonds. Some impurities may remain in the lignin structure, which may partially block the functionalization capacity; for instance, the presence of sulfur from the Na_2S may limit the functionalization process due to its multivalent nature. As the kraft process was carried out at high-temperature and high-alkaline medium, aryl bonds break down and form a significant number of phenolic hydroxyls. Besides, the delignification process is carried out in an oxidative environment, which may form catechol, quinone structures, and carboxylic groups. Kraft lignin also contains trace amounts of carbohydrate and ash, as well as aliphatic thiol groups, which result in high polydispersity and low molecular weight [14–16].

Lignosulfonate Lignin

Lignosulfonate lignin is produced as a by-product during the sulfite pulping process of cellulose, where a mixture of alkaline earth metal sulfite delignifies the lignocellulose materials. Lignin produced in this process is degraded by acid hydrolysis and becomes a sulfonated compound, which enables it to harness various functional groups. Such functionalization contributes to various features of the lignin, for instance, water solubility, distinctive colloidal properties, increased molecular weight, and higher ash content (4–8%) than kraft lignin [17]. It has broad applications as binders, surfactants, dispersants, adhesives, additives for cement, suspensions, stabilizing agents, and so on [18].

2.1.2 Sulfur-Free Lignin

Soda Lignin

This process is similar to the kraft lignin process, except the delignification process is carried out without using sodium sulfide (Na_2S). As the process is sulfur-free, soda lignin is edible and used for animal nutrition, phenolic resins, dispersants, polymer synthesis, etc. However, only NaOH is used throughout the process, which reduces production efficiency. Hence, the soda lignin process is not very suitable for industrial production [18].

Organosolv Lignin

In the organosolv process, organic solvents (ethanol, methanol, acetone, acetic acid, formic acid, etc.) are used to isolate the lignin from the lignocellulosic biomass. The primary advantages of this process are its sulfur-free environment and the reusability of its solvents [19]. In this process, the lignocellulosic biomass is treated by any of the aforementioned organic solvents at 120–220 °C and 1–4 atm pressure. This process dissolves lignin (keeping cellulose intact), which can be purified over 90% through a series of filtration, solvent evaporation, and washing processes [20]. Finally, the produced organosolv lignin is recovered from the solution via precipitation, which requires adjusting the pH, concentration, and temperature [21]. Two distinct attributes of organosolv lignin are high hydrophobicity and 36–56% lower molecular weight from the lignin source. Besides, the lignin produced also has a homogeneous structure, high purity and chemical reactivity, and zero toxicity [9, 17].

2.2 *Lignin Modifications and Blends*

Lignin is composed of aromatic rings, oriented in all three dimensions, forming highly crosslinked networks (Fig. 1b) with limited ability of bending and stretching without being ruptured. However, to generate more value-added products, lignin is often subjected to various chemical modifications that add chemically active sites (Fig. 2a). Lignin also possesses many aliphatic and phenolic hydroxyl groups that can be functionalized (Fig. 2b) to form highly reactive polyol derivatives that are useful across various applications, such as biomedical, energy, and flexible electronics [22].

2.2.1 Generating Chemically Active Sites

Chemical modifications can create chemically active sites on lignin structure, which will open new opportunities to use lignin efficiently. Usually, during these modifications, the respective functional groups form covalent bonds to the C5 position of the aromatic ring of guaiacyl (G) and syringyl (S) units [23]. Chemical modifications such as hydroxyalkylation, amination, sulfomethylation, sulfonation, and nitration are well-known (Fig. 2a). Hydroxyalkylation increases the surface area and adsorption capacity of lignin, along with allowing more chemical groups to interact with the depolymerized kraft lignin components [24]. Amination combines amine functional groups with lignin derivatives. Cationic lignin can be obtained from the amination process in an alkaline environment [25]. Methylene sulfonate groups can also be added to lignin via sulfomethylation with sodium sulfite and formaldehyde. This produces sulfonated polyphenols, such as sulfomethylated tannic acids [22]. Sulfonation of lignin brings significant changes in properties by changing molecular weight, charge density, and solubility, leveraging the opportunities for surface modification, adsorption, etc. Sulfonated lignin could be used as a coagulant for dye

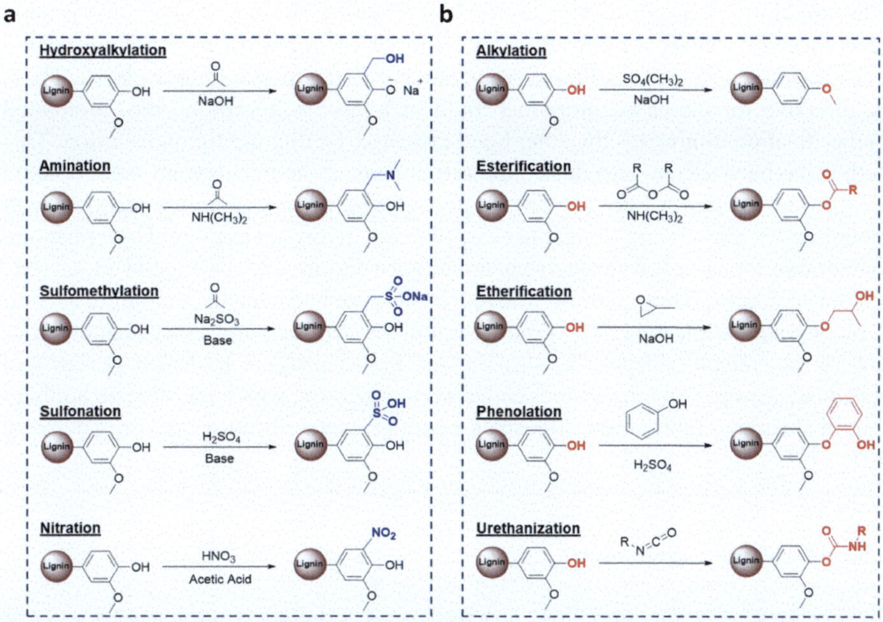

Fig. 2 **a** Chemical reactions of lignin for generating new active sites: hydroxyalkylation, amination, sulfomethylation, sulfonation, and nitration. **b** Reactions to functionalize the hydroxyl groups through alkylation, esterification, etherification, phenolation, and urethanization. Reproduced under the terms of the creative commons attribution license [22]

removal [26]. In the nitration process to create active sites in lignin, nitrating agents are commonly used, i.e., nitric acid with acetic anhydride, sulfuric acid, and acetic acid. The nitration of lignin can slightly increase the crosslinking sites [27].

2.2.2 Functionalization of Hydroxyl Groups

The phenolic and aliphatic hydroxyl groups on lignin structure could be functionalized and create polyol derivatives, which unlock various applications (Fig. 2b). Alkylation of lignin with 1,6-dibromohexane has been shown to improve photoinitiation effects, promote crosslinking, and significantly increase molecular weight [25, 28]. On the other hand, the esterification process contributes to the ring-opening of cyclic esters as well as polymerization with carboxylic acid chloride and dicarboxylic acids. This process has been used to synthesize high capacitance and stable carbon fibers from organosolv lignin. A greener esterification technique was also developed by functionalization of kraft lignin with organic acids, which produced a polymer blend with a lower glass transition temperature. Esterification of kraft lignin with methacrylic anhydride was also carried out to accommodate into poly(ethylene glycol)diacrylate (PEGDA) matrices, resulting in improved mechanical strength, light shielding, and antioxidant performance, which are essential for 3D-printed structures [29]. Next, the etherification methods of lignin involve polymerization with alkylene oxides or epichlorohydrin, and solvolysis with ethylene glycol. These are effective for increased monomer yield of lignin. Furthermore, diol pretreated α-etherification minimizes the ether bond cleavage, facilitating lignin isolation. This method has been used to introduce photoinitiating macromolecules and water-soluble PEG groups to lignin [30, 31]. In the phenolation process, lignin's reaction with phenol under acidic conditions, raises its phenolic hydroxyl content. High phenolic content can be achieved through phenolation of lignin with pyrogallol in acetone and sulfuric acid. These can be utilized in polyurethane foams, coatings, as well as biodegradable plastics [32]. Finally, lignin's reactions with diisocyanate and/or diol, or isocyanate and polyol, forms a link between lignin hydroxyl groups and isocyanate groups. This process is called urethanization, which has diverse applications in elastomer fabrication, adhesives, coatings, water filtration, and drug delivery [33].

2.2.3 Grafting Copolymers

To incorporate new functionalities in lignin, graft copolymerization can be a viable technique. It improves lignin's homogeneity, mechanical properties, and 3D printability. In lignin, phenolic and aliphatic hydroxyl groups are omnipresent, which creates the reactive sites for docking diverse external copolymers through various techniques, such as ring-opening polymerization (ROP), reversible addition–fragmentation chain transfer polymerization (RAFT), atom transfer radical polymerization (ATRP), etc. [34, 35]. With these techniques, suitable copolymers for various applications could be tailored with the lignin structure through the reactions we have discussed earlier: alkylation, esterification, amination, etc. Grafting copolymers on the lignin structure significantly increases molecular weight, along with changes in mechanical, thermal, and rheological behavior [36, 37].

3 3D Printing Techniques Using Lignin

3D printing techniques are increasingly utilized for rapid prototyping and the fabrication of complex components without the need for expensive molds or tooling. It offers significant advantages, including design flexibility, near-net shape fabrication, and minimal material waste, making them valuable in various industries, such as energy, aerospace, automobile, and biomedical engineering. Researchers have combined lignin with appropriate chemical formulations to make it compatible with various 3D printing techniques [38–40]. This section describes how researchers have successfully incorporated lignin into filaments, resins, and inks for 3D printing using FFF, SLA, and DIW, respectively, addressing the challenges of printability with each technique.

3.1 Fused Filament Fabrication (FFF) of Lignin

In recent years, researchers have developed lignin-based composite filaments for 3D printing via FFF. In this process, thermoplastic filaments are melted in an extrusion hot end and deposited layer by layer to construct intricate three-dimensional structures. A schematic representation of the FFF process is illustrated in Fig. 3a.

One of the challenges in incorporating lignin into FFF filaments is the brittleness of biopolymers due to their complex aromatic molecular structure. Lignin softening occurs between 90–180°C without exhibiting a sharp melting point. Also, above 180°C, the evaporation of volatile compounds compromises print quality and mechanical stability [42, 43]. To counteract this, extrusion parameters must be pre-optimized to avoid uncontrolled softening. Such conditions could be reduced by incorporating other thermoplastic materials, such as nylon and ABS, with the lignin as a composite, (Fig. 4) [44]. Nguyen et al. [42] blended lignin at 40 wt% and 60 wt%, both with and without carbon fiber (CF) reinforcements, using ABS and nylon as matrices. The printed structures of lignin-modified nylon composites demonstrated enhanced stiffness and tensile strength at room temperature. Additionally, the increased tensile strength confirmed the potential of lignin-based filaments as viable reinforcement materials for FFF 3D printing applications.

The successful integration of lignin into FFF 3D printing would not only enhance its industrial utilization but also contribute to sustainable development of biocompatible polymeric composites. Moreover, previous studies have demonstrated that lignin can reinforce the mechanical properties of biopolymers, highlighting its potential as a functional additive for next-generation sustainable manufacturing [45, 46]. Kai et al. [47] combined poly(3-hydroxybutyrate) (PHB), a brittle biopolymer, with 20 wt% lignin nanoparticles to produce PHB/lignin composite filaments. The printed PHB/lignin composite exhibited a 73% increase in tensile strength and 267% greater plastic deformation at necking compared to those printed with pure PHB. However, there is a limitation to using excessive lignin nanoparticles as the clustering and heterogeneity of lignin can deteriorate the mechanical strength of the composites.

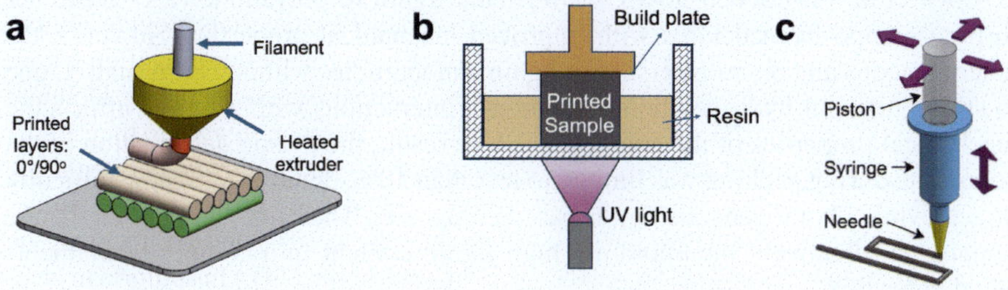

Fig. 3 Conventional 3D printing technologies suitable for printing lignin biopolymer and lignin-based composites: **a** Fused filament fabrication (FFF), **b** Stereolithography (SLA), and **c** Direct ink writing (DIW). Reproduced with permission from Coffigniez et al. [41]

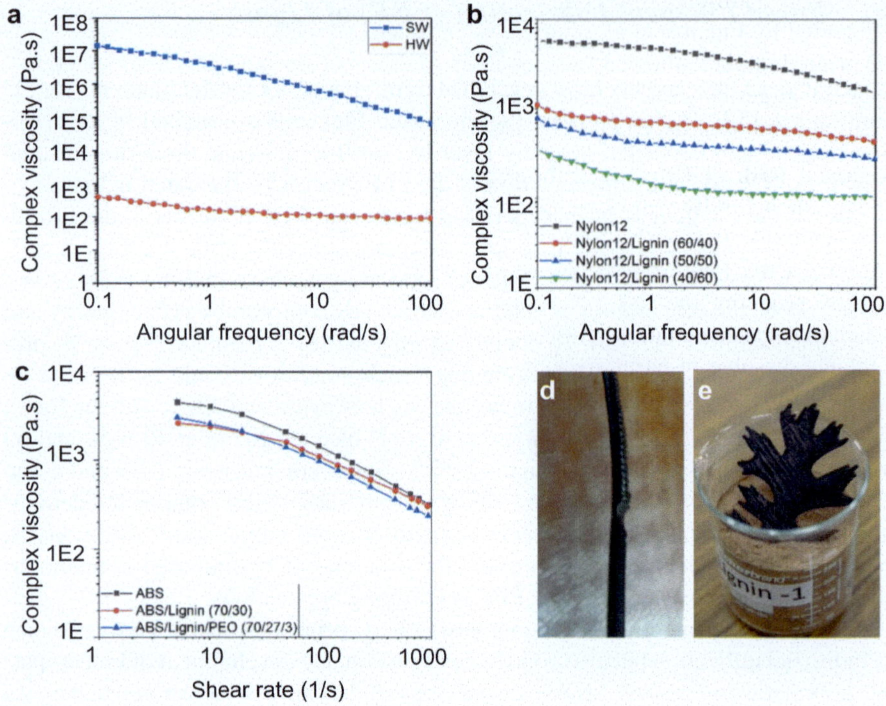

Fig. 4 Rheological properties and 3D printability of lignin derivatives via FFF, assessed through layer-by-layer melt deposition, based on zero shear viscosity, shear-thinning behavior, and room-temperature stiffness and toughness of the extrudable material. **a** Organosolv hardwood (HW) lignin has lower viscosity than Kraft softwood (SW) lignin at 190 °C; **b** HW lignin reduces nylon12 viscosity at 210 °C, with higher lignin content further lowering melt viscosity. **c** The ABS matrix shows minimal viscosity reduction with 30 wt% HW lignin. **d** Buckling-induced failure in soft compositions that limit printability; **e** 3D-printed oak leaf. Reproduced under the terms of creative commons attribution license [6]

Chen et al. [48] added 50 wt% enzymatic lignin to polyamide (PA) elastomer to print composite structures with improved mechanical properties. However, the heterogeneous and poor dispersion of lignin nanoparticles within the PA matrix, due to strong inherent hydrogen bonding and π-π interactions within lignin, affects the mechanical properties of the composites. As a result, the plastic deformation of the composite during necking was limited to less than 10%, demonstrating the difficulty of achieving a homogeneous lignin-based composite. Therefore, a major challenge remains in effectively incorporating high lignin content into composite filaments suitable for FFF.

Lignin has also been combined with thermoplastic polyurethane (TPU), an elastomer known for its excellent fatigue resistance, wear resistance, and thermoplasticity, making it well-suited for 3D printing via FFF. Researchers have found that the synergistic effect of TPU with lignin can increase the ductility, making it suitable

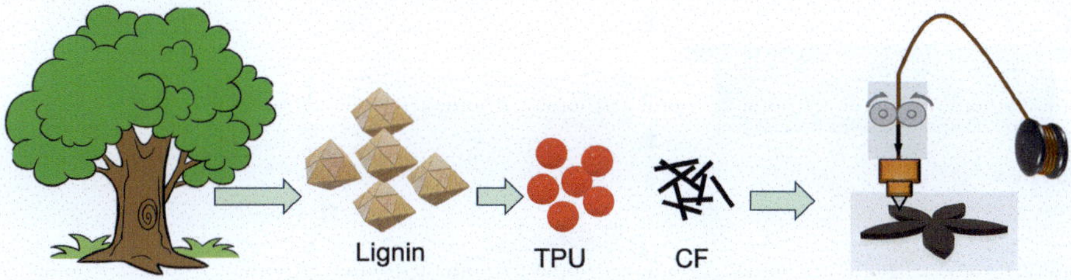

Fig. 5 Schematic representation of fused filament fabrication 3D printing using lignin, thermoplastic polyurethane (TPU), and carbon fiber (CF) composite. Reproduced with permission from X. Zhou et al. [49]

for FFF with enhanced mechanical properties [50–54]. In one study, a cost-effective and sustainable composite filament was developed by Zhou et al. [49] with TPU elastomer and chopped CF as the reinforcements, and lignin as a matrix (Fig. 5). In this formulation, TPU elastomer was used to enhance the elasticity and plasticity of the lignin while also modifying the rheology of the lignin to make it printable. On the other hand, CF provides excellent mechanical properties for the TPU/lignin composites with improved interlayer adhesion. The FFF printed composite filaments exhibited stable and continuous printing performance, with excellent thermal stability and increased mechanical strengths. Overall, this unveils a new horizon for developing sustainable, cost-effective, and biocompatible lignin-based FFF 3D printing filaments.

3.2 *Stereolithography (SLA) of Lignin*

SLA 3D printing process uses a vat containing liquid photopolymer resin that undergoes polymerization to form crosslink networks at specific locations, where UV-light exposure triggers the curing reaction (Fig. 6a). The localized crosslink formation enables precise 3D printing of intricate structures with excellent surface finishes and high resolution [55]. Efficient polymerization through photo-crosslinking requires an appropriate resin formulation for the SLA process. Functionalization and epoxidation of various unsaturated moieties, such as carboxyl, epoxy, and hydroxyl groups, through reactions with glycidyl methacrylate, methacrylic acid, or methacrylic anhydride are some examples [55]. To promote crosslinking and improve photoinitiation effects in a lignin blend, alkylation with 1,6-dibromohexane has been used. Researchers found that this blend significantly increased the molecular weight of lignin by 1643%, which is essential for photoinitiation and effective SLA 3D printing [25, 28]. In a recent study, lignin biopolymer was functionalized with methacrylate, rendering it photo-curable for SLA 3D printing. The modified lignin was incorporated into a commercially available polymeric resin at concentrations of up to 15 wt% and utilized in the SLA process. The results exhibited that the plasticizing property of the modified lignin led to a reduction in stiffness, increasing ductility 4 times [56].

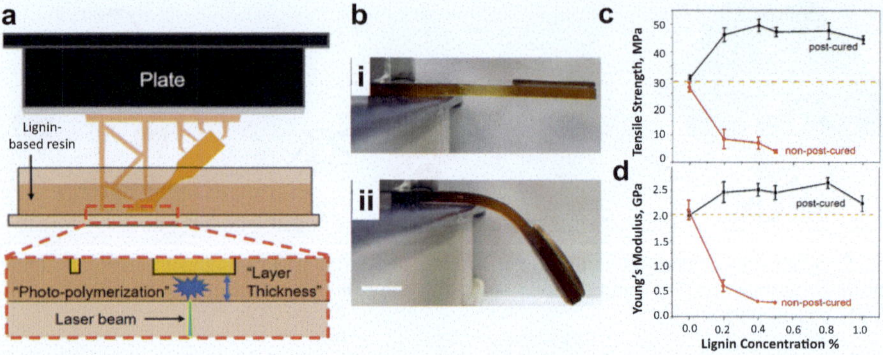

Fig. 6 Lignin-reinforced composite resin for SLA 3D printing. **a** Schematic representation of SLA 3D printing technology utilizing lignin composite resin [58]. **b** Illustration showing the effect of lignin concentration on mechanical stiffness: (i) 0.2 wt% lignin, (ii) 1.0 wt% lignin, demonstrating that higher concentration of lignin causes lower stiffness of the printed samples [58]. **c** Tensile strengths of the SLA printed lignin composite samples before thermal curing (red) and after thermal curing (black), demonstrating the dramatic increase in tensile strength after curing with different lignin concentrations in the composite resin [58]. **d** Young's Moduli of the SLA printed lignin composite samples before thermal curing (red) and after thermal curing (black), demonstrating the increase in mechanical stiffness after curing with increasing lignin concentrations in the composite resin [58]. Reproduced with permission from Zhang et al. [58]

Vanillin is one of the derivatives of lignin biopolymer, which has also been explored for SLA 3D printing. Initially, vanillin was functionalized through methacrylation and then equimolarly dispersed into glycerol dimethacrylate. Although the resulting solution occupied 35% lignin, the 3D-printed architectures exhibited excellent mechanical and thermal properties, including a high material stiffness of 4900 MPa and high glass transition temperature of 153 °C. This highlights the potential of lignin and lignin-derived biopolymers in SLA, paving the path toward environmentally benign photo-curable resins [57]. Zhang et al. [58] also developed an SLA resin formulation combining lignin with methacrylate resin at varying concentrations (0.2, 0.4, 0.5, 0.8, and 1.0 wt%). The study demonstrated that lignin reinforcement significantly enhanced mechanical performance. Post-curing led to a 46–64% increase in tensile strength (Fig. 6c) and a 13–37% improvement in Young's modulus (Fig. 6d) compared to the lignin-free control sample. Notably, the optimal lignin concentration was 0.4 wt%, yielding a maximum tensile strength of 49.0 MPa, which is a 60% improvement over the tensile strength of the control sample. Higher lignin content affected the optical properties of the printed parts. Increasing the lignin concentration darkened the material and resulted in a softer texture (Fig. 6b).

3.3 *Direct Ink Writing (DIW) of Lignin*

DIW is one of the most versatile 3D printing techniques for almost any material if it can be modified to attain the appropriate rheology [59]. This technique can be a better process for 3D printing of lignin, as DIW operates at ambient conditions, unlike FFF which requires elevated temperatures leading to lignin softening [60]. However, lignin alone exhibits poor rheological properties when directly mixed with water or organic solvents, leading to inconsistent extrusion and nozzle clogging, affecting the print fidelity. Rheology modifiers, such as cellulose nanocrystals (CNC), nanoclay, methylcellulose, and colloidal gels, can induce rheology conducive to DIW. To incorporate lignin in DIW, the lignin-based viscoelastic ink should exhibit shear-thinning behavior upon extrusion for smooth layer deposition, while also demonstrating shape retention capability. The morphology and molecular structure of lignin influence the material characteristics, particularly the viscoelasticity necessary for the appropriate formulations of DIW 3D printing [7].

Zhang et al. [61] formulated a hydrogel composed of colloidal lignin (CL), alginate (A), cellulose nanofibrils (CNF), and Ca^{2+} for crosslink network formation. Figure 7a shows the schematic representation for DIW 3D printing utilizing the A-CNF-CL composite hydrogel. They found that the viscosity of the A-CNF-CL composite hydrogel decreases with 1.0 wt% CL but after that increases with increasing CL concentrations (5 to 25 wt%). Figure 7b-iv and b-v are the DIW-printed scaffolds of A-CNF-CL composite hydrogels at CL concentrations of 10 wt% and 25 wt%, respectively, indicating the shape retention capability of the hydrogels after being deposited. Notably, even at a low concentration of 0.5 wt%, lignin nanoparticles were effective in enhancing the hydrogel's shape retention capability and providing antioxidant properties.

Raman et al. [62] performed a study on the flow rheology of a lignin-based formulation with graphene oxide (GO), and they used water-soluble lignin instead of insoluble lignin (Fig. 7b-i). The results show that the viscosity of lignin/GO inks decreases with lignin content, and the pure lignin solution shows negligible viscosity change with shear stress. The GO content in the composite inks significantly influences the flow rheology, and the composite ink shows higher printability with GO concentrations. At lower shear stress, these inks exhibit a significantly higher storage modulus ($G^{/}$) than loss modulus ($G^{//}$), promoting the formation of stable structures. Figure 7b-ii and b-iii are the DIW-printed samples of GO/lignin composite inks after they have been freeze-dried and thermally dried, respectively. Overall, this unique combination of lignin and GO contributed to printability, shape retention capability, and material stiffness of the DIW-printed parts.

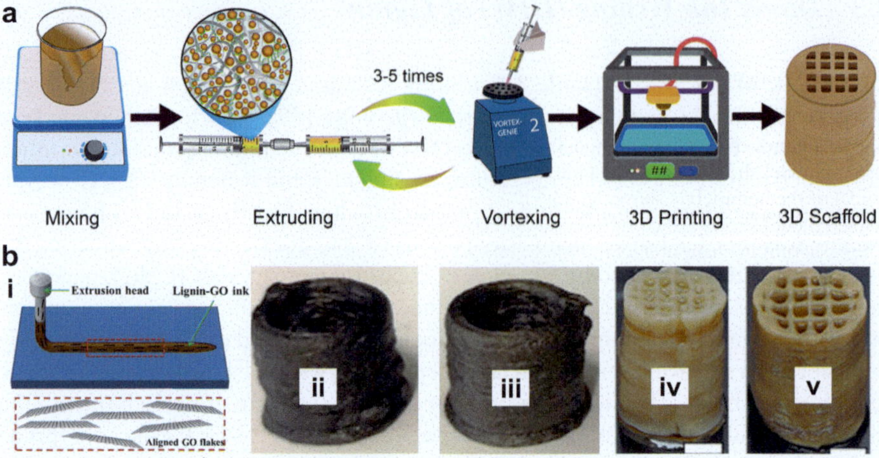

Fig. 7 DIW 3D printing with hydrogels of lignin-based composites. **a** Schematic of colloidal lignin particles and cellulose nanofibril-alginate hydrogel preparation for DIW 3D printing [61]. **b** DIW-printed lignin scaffolds: (i) Schematic showing graphene oxide (GO) flake alignment during extrusion and after deposition[62], (ii) Freeze-dried lignin/GO scaffold [62], (iii) Thermally dried lignin/GO scaffold [62], (iv) DIW-printed A-CNF-CL scaffold at CL concentration of 10 wt% [61], and (v) DIW-printed A-CNF-CL scaffold at CL concentration of 25 wt% [61], highlighting the versatility of the DIW 3D printing process in fabricating lignin-based composite scaffolds with tailorable mechanical properties. Reproduced under the terms of creative commons attribution license [61], and reproduced with permission from Roman et al. [62]

4 3D Printing of Wood-Like Materials from Lignin

Natural wood has long been used as a fundamental material for many different applications, including building construction, furniture production, and architectural design [63, 64]. Historically, wood shaping has depended on subtractive manufacturing methods. Nevertheless, this approach frequently produces a large amount of waste. On the contrary, AM produces much less material waste and also gives greater design freedom, which has encouraged recent researchers to make several attempts at 3D printing wood-like materials. Naturally, the closest that researchers have gotten to mimicking natural wood is by utilizing the fundamental building block of natural wood—lignin [65–68]. They achieved this by creating a mixture of lignin and cellulose, which is the other basic component of natural wood. The ratios of this mixture are carefully tuned, and sometimes binders or rheology modifiers are added to attain the rheology requirements of DIW.

Combining lignin and cellulose for AM is a relatively new field. Thomas Liebrand, in his investigation into 3D-printed fiber-reinforced lignin, conducted experiments with a combination of kraft lignin and bleached kraft cellulose sheets that were pulped, mixed with demineralized water and acetone in varying ratios to develop an extrudable material [66]. However, the hazardous nature and flammable characteristics of acetone suggested that any future advancements in the material would necessitate the investigation of different binding agents [67].

Bierach et al. worked on the same "Wood Without Trees" research line at TU Delft, following up on Liebrand's work [67]. Their goal was to advance the development of a wood-like feedstock, identify a suitable binding agent, and utilize it for prototype fabrication. After assessing 12 different bio-based and synthetic binders, they achieved the best rheology with methylcellulose and wood glue. Yet, in prototype fabrication, they found that while curvilinear walls with a dense infill without any overhangs were stable, linear structures made with single-line extrusions were prone to buckling and collapse under their own weight. Furthermore, the printed structures showed uneven shrinkage during drying, leading to large shape changes, and the printing resolution was also low, thus leading to thick layers and large interlayer separations.

At the Maksud Research Group, we mimicked the composition of natural wood by mixing 25 wt% organosolv hardwood lignin with cellulose nanocrystals and nanofibers [65] (Fig. 8). This was the primary focus of Thakur's Ph.D. thesis [68], and continues to be one of the primary directions of research from our current group. We formulated an additive-free water-based ink that achieved appropriate rheology for DIW. This enabled us to print highly intricate architected wood structures at the highest resolution so far reported using 260 μm diameter nozzles with Hyrel 3D printers (Hyrel Engine HR & System 30 M). A room-temperature cold flow high torque syringe head (SDS-60XT Extruder) is used to extrude the ink from a 60 mL Luer-lock syringe. Smooth-flow tapered tips (Nordson EFD) were utilized to prevent clogging and printing discontinuities. The ink was printed on a silicone substrate to facilitate the detachment of the printed object from the build plate and prevent adhesion between the wood structure and the build plate. The geometry and additional parameters, such as extrusion width, printing speed, and layer height, were used to generate the G-code script using the PrusaSlicer software. The dimensions of the samples printed were 15 × 5 × 5 mm and 35 × 5 × 5 mm for compression tests and flexure tests, respectively. To retain the architectures and dimensions, the printed parts are freeze-dried, since drying in ambient conditions leads to uncontrolled deformation and bending. The resulting printed structures, after post-processing, closely resembled the visual, textural, olfactory, and macro-anisotropic properties, including mechanical properties, of natural wood [65] (Fig. 9a). The heat treatment was performed at 180 °C, which is above the softening point of lignin (~140 °C). Therefore, during this heat treatment, the lignin can flow inside the structure and bind the other components, cellulose nanocrystals, and nanofibers, leading to a more resilient structure. We also showed that the process is recyclable, as we were able to grind down a printed structure, re-add the water, to reproduce the ink, which we could utilize to 3D print a completely new structure.

Recently, we have further developed this work by introducing natural fibers, flax and hemp, into the wood-ink formulation, which significantly improved the mechanical properties [68]. We added 1 wt. % of chopped fibers, ~ 5mm in length, to the formulation for fiber-reinforced 3D-printed wood. Next, they are mixed in a Thinky

Fig. 8 Three-dimensional printing of wood from its fundamental building blocks, lignin, and cellulose. Reproduced under the terms of creative commons attribution license [65]

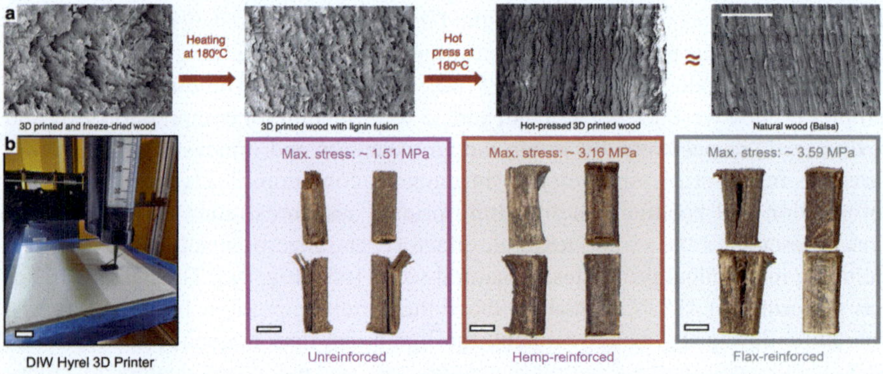

Fig. 9 **a** SEM micrographs of 3D-printed wood at each post-processing, showing the hot-pressed 3D-printed wood mimicking the natural balsa wood microstructure [65]. **b** Direct ink writing (DIW) Hyrel 3D printer used for 3D printing wood and the unreinforced, hemp-reinforced, and flax-reinforced 3D-printed wood samples after compression tests, from left to right, respectively [68]. First scale bar is 15 mm, the remaining scale bars are 5 mm

Planetary Centrifugal Mixer (model AR 230), three cycles of mixing at 2000 rpm for five minutes, followed by defoaming for one minute at 2200 rpm. Then the mixture is loaded into the syringe for printing. (Fig. 9b) Both hemp and flax fibers substantially improved compressive strength by 86% and 81%, respectively, over 3D-printed unreinforced wood. In Fig. 9, the unreinforced 3D-printed wood shows a split failure at ~ 1.51 MPa with layers separating due to the interfaces formed during the layer-by-layer printing process. Although the hemp-reinforced printed wood shows a split failure mode too, it withstands a higher stress of ~ 3.16 MPa, and the printed layers remain together. The flax-reinforced 3D-printed wood shows a wedge-split failure withstanding a higher stress level of 3.59 MPa. The increase in strength can be attributed to the high tensile strengths of the natural fibers.

3D printing of wood from lignin is a fairly new field which is quickly gaining traction in recent years. It has the potential to pave the way toward sustainable AM of architected wood structures, especially for the future of wood construction in remote areas, including space.

5 Applications of 3D-Printed Lignin-Based Structures

Up to this point, the potential of lignin, the second most abundant biopolymer, is understandable; however, due to the lack of available technology to properly valorize it, nearly 50 million tons are wasted every year. Intensive research is going on to repurpose lignin in the field of additive manufacturing, because of its bio-based origin, renewability, and potential to replace hydrocarbon-based polymers. Some of the emerging applications of lignin in the fields of construction, furniture, and biomedical sectors are discussed here, along with the future challenges and opportunities in lignin-based 3D printing.

The mechanical properties and thermal stability of lignin make it a promising candidate for construction and furniture applications. Lignin-based blends/composites can be used in large-scale additive manufacturing processes to manufacture eco-friendly construction materials. Combining with hydrocarbon-based thermoplastics, such as ABS or PP, lignin improves the structural stability of the fabricated structure while reducing their carbon footprints [69]. Lignin-based composites have also been investigated for mimicking wooden furniture. Researchers have developed bio-filaments by blending lignin, cellulose, and hemicellulose, which are suitable for FFF. The appearance and texture of these materials resemble the original wood while offering improved durability and water resistance. These innovations can pave the way toward sustainable household and construction utility manufacturing as well as reducing dependency on deforestation and hydrocarbon-based polymers [70].

Lignin also has antioxidant, antimicrobial, and biocompatible properties, which are the major requirements for a material to be used in clinical purpose. We have discussed the 3D printability of lignin combining with other biodegradable polymers, i.e., polylactic acid (PLA) or polycaprolactone (PCL), which can be utilized for 3D-printed scaffolds in tissue engineering. Furthermore, their biocompatibility and tunable biodegradability, made them ideal for regenerative medicine applications [71]. Lignin-based hydrogels are also investigated for wound dressings, drug delivery systems and implantable devices. The phenolic groups present in lignin contribute to free-radical scavenging, which can protect tissues from oxidative stress and inflammation [72]. 3D bioprinting with lignin nanoparticle mixed bio-ink has enabled the fabrication of patient-specific prosthetics and implants with improved mechanical properties. These applications align with the recent trend of personalized medicine, where bio-based materials play a vital role in developing sustainable and patient-friendly solutions.

6 Conclusion

The advancement of 3D printing technology has provided new avenues for valorizing lignin, a renewable and sustainable biopolymer. As discussed throughout this chapter, integrating lignin into 3D printing processes presents both challenges and opportunities. While lignin's complex structure and thermal behavior pose difficulties in processing, innovations in material formulation, such as blending with thermoplastics, chemical modifications, and composite approaches, have demonstrated promising results. However, further advancements in processing techniques and compatibility optimization are required to fully unlock lignin's potential in additive manufacturing. Future studies should focus on refining lignin extraction and modification techniques to improve its printability and mechanical performance. Additionally, a deeper understanding of lignin's interactions with polymers at the molecular level will enable the development of more efficient and scalable 3D printing strategies. Collaborative efforts between researchers, industry stakeholders, and policymakers will be essential in promoting the adoption of lignin-based 3D printing materials, reducing dependence on fossil-based polymers, and advancing the circular economy. Overall, while challenges remain, the progress in lignin-based 3D printing demonstrates its potential as a viable and sustainable alternative to traditional synthetic materials. With continued research and technological advancements, lignin could play a significant role in the future of additive manufacturing, contributing to more eco-friendly and resource-efficient production systems.

References

1. Ralph, J., Lapierre, C., Boerjan, W.: Lignin structure and its engineering. Curr. Opin. Biotechnol. **56**, 240–249 (2019). https://doi.org/10.1016/j.copbio.2019.02.019
2. Cho, D.-W., Yoon, K., Ahn, Y., Sun, Y., Tsang, D.C.W., Hou, D., Ok, Y.S., Song, H.: Fabrication and environmental applications of multifunctional mixed metal-biochar composites (MMBC) from red mud and lignin wastes. J. Hazard. Mater. **374**, 412–419 (2019). https://doi.org/10.1016/j.jhazmat.2019.04.071

3. Sun, R.: Lignin source and structural characterization. Chemsuschem **13**, 4385–4393 (2020). https://doi.org/10.1002/cssc.202001324
4. Glasser, W.G.: About making lignin great again—some lessons from the past. Front. Chem. **7**, 565 (2019)
5. Bajwa, D.S., Pourhashem, G., Ullah, A.H., Bajwa, S.G.: A concise review of current lignin production, applications, products and their environmental impact. Ind. Crops Prod. **139**, 111526 (2019)
6. Ebers, L.-S., Arya, A., Bowland, C.C., Glasser, W.G., Chmely, S.C., Naskar, A.K., Laborie, M.-P.: 3D printing of lignin: challenges, opportunities and roads onward. Biopolymers **112**, e23431 (2021). https://doi.org/10.1002/bip.23431
7. Jiang, B., Yao, Y., Liang, Z., Gao, J., Chen, G., Xia, Q., Mi, R., Jiao, M., Wang, X., Hu, L.: Lignin-based direct ink printed structural scaffolds. Small **16**, 1907212 (2020). https://doi.org/10.1002/smll.201907212
8. Gleuwitz, F.R., Sivasankarapillai, G., Siqueira, G., Friedrich, C., Laborie, M.-P.G.: Lignin in bio-based liquid crystalline network material with potential for direct ink writing. ACS Appl. Bio Mater. **3**, 6049–6058 (2020). https://doi.org/10.1021/acsabm.0c00661
9. José Borges Gomes, F., De Souza, R.E., Brito, E.O., Costa Lelis, R.C.: A review on lignin sources and uses. J. Appl. Biotechnol. Bioeng. 100–105 (2020). https://doi.org/10.15406/jabb.2020.07.00222
10. Nishimura, H., Kamiya, A., Nagata, T., Katahira, M., Watanabe, T.: Direct evidence for α ether linkage between lignin and carbohydrates in wood cell walls. Sci. Rep. **8**, 6538 (2018). https://doi.org/10.1038/s41598-018-24328-9
11. Rodrigues Mota, T. et al.: Plant cell wall composition and enzymatic deconstruction. AIMS Bioengineering **5**, 63–77 (2018)
12. Vishtal, A., Kraslawski, A.: Challenges in industrial applications of technical lignins. BioResources. **6** (2011)
13. Capanema, E.A., Balakshin, M.: Plantrose lignins: a new type of technical lignins. In: Proceedings of the 18th international symposium on wood, fiber and pulping chemistry, Vienna, Austria, September (2015)
14. Lin, X., Zhou, M., Wang, S., Lou, H., Yang, D., Qiu, X.: Synthesis, structure, and dispersion property of a novel lignin-based polyoxyethylene ether from kraft lignin and poly (ethylene glycol). ACS Sustain. Chem. Eng. **2**, 1902–1909 (2014)
15. Asgari, F., Argyropoulos, D.S.: Fundamentals of oxygen delignification. Part II. Functional group formation/elimination in residual kraft lignin. Can. J. Chem. 76, 1606–1615 (1998)
16. Luong, N.D., Binh, N.T.T., Duong, L.D., Kim, D.O., Kim, D.-S., Lee, S.H., Kim, B.J., Lee, Y.S., Nam, J.-D.: An eco-friendly and efficient route of lignin extraction from black liquor and a lignin-based copolyester synthesis. Polym. Bull. **68**, 879–890 (2012)
17. Kumar, A., Kumar, J., Bhaskar, T.: Others: Utilization of lignin: a sustainable and eco-friendly approach. J. Energy Inst. **93**, 235–271 (2020)
18. de Almeida, D.P., Gomide, J.L.: Anthraquinone and surfactant effect on soda pulping. O Pap. **74**, 53–56 (2013)
19. Yáñez-S, M., Matsuhiro, B., Nuñez, C., Pan, S., Hubbell, C.A., Sannigrahi, P., Ragauskas, A.J.: Physicochemical characterization of ethanol organosolv lignin (EOL) from eucalyptus globulus: effect of extraction conditions on the molecular structure. Polym. Degrad. Stab. **110**, 184–194 (2014). https://doi.org/10.1016/j.polymdegradstab.2014.08.026
20. Lora, J.H., Glasser, W.G.: Recent industrial applications of lignin: a sustainable alternative to nonrenewable materials. J. Polym. Environ. **10**, 39–48 (2002). https://doi.org/10.1023/A:1021070006895
21. Vázquez, G., Antorrena, G., González, J., Freire, S.: The influence of pulping conditions on the structure of acetosolv eucalyptus lignins. J. Wood Chem. Technol. **17**, 147–162 (1997). https://doi.org/10.1080/02773819708003124
22. Wu, T., Sugiarto, S., Yang, R., Sathasivam, T., Weerasinghe, U.A., Chee, P.L., Yap, O., Nyström, G., Kai, D.: From 3D to 4D printing of lignin towards green materials and sustainable manufacturing. Mater. Horiz. (2025). https://doi.org/10.1039/D4MH01680G

23. Lee, Y., Tran, M.H., Lee, E.Y.: Hydroxymethylation of technical lignins obtained from different pretreatments for preparation of high-performance rigid polyurethane foam. Eur. J. Wood Wood Prod. **80**, 1225–1234 (2022). https://doi.org/10.1007/s00107-022-01845-z
24. Vaz, S., Jr., Salvador, C.E.D.M.: Innovative structural modification process of Kraft lignin using continuous-flow regime. Sustain. Chem. Pharm. **36**, 101273 (2023). https://doi.org/10.1016/j.scp.2023.101273
25. Jiang, X., Tian, Z., Ji, X., Ma, H., Yang, G., He, M., Dai, L., Xu, T., Si, C.: Alkylation modification for lignin color reduction and molecular weight adjustment. Int. J. Biol. Macromol. **201**, 400–410 (2022). https://doi.org/10.1016/j.ijbiomac.2021.12.162
26. Hu, S., Zhang, T., Jiang, B., Huang, C., Wei, W., Wu, W., Jin, Y.: Achieving high enzymatic hydrolysis sugar yield of sodium hydroxide-pretreated wheat straw with a low cellulase dosage by adding sulfomethylated tannic acid. Bioresour. Technol. **384**, 129276 (2023). https://doi.org/10.1016/j.biortech.2023.129276
27. Abbas, A., Wang, Z., Zhang, Y., Peng, P., She, D.: Lignin-based controlled release fertilizers: a review. Int. J. Biol. Macromol. **222**, 1801–1817 (2022). https://doi.org/10.1016/j.ijbiomac.2022.09.265
28. Zhang, X., Keck, S., Qi, Y., Baudis, S., Zhao, Y.: Study on modified dealkaline lignin as visible light macromolecular photoinitiator for 3D printing. ACS Sustain. Chem. Eng. acssuschemeng (2020). https://doi.org/10.1021/acssuschemeng.0c03587
29. Park, H.W., Ahn, H.-J., Roh, K.C.: Recent advances in preparation and supercapacitor applications of lignin-derived porous carbon: a review. J. Electrochem. Sci. Technol. **15**, 111–131 (2024). https://doi.org/10.33961/jecst.2023.00892
30. Figueiredo, P., Lintinen, K., Hirvonen, J.T., Kostiainen, M.A., Santos, H.A.: Properties and chemical modifications of lignin: towards lignin-based nanomaterials for biomedical applications. Prog. Mater. Sci. **93**, 233–269 (2018). https://doi.org/10.1016/j.pmatsci.2017.12.001
31. Dong, C., Meng, X., Leu, S.-Y., Xu, L., Wu, Z., Cravotto, G., Fang, Z.: Enhancing α-etherification of lignin in eucalyptus diol pretreatment to improve lignin monomer production. Ind. Crops Prod. **185**, 115130 (2022). https://doi.org/10.1016/j.indcrop.2022.115130
32. Zhang, Z., Liu, Q., Gao, T., Qiao, C., Yao, J., Zhang, C.: Phenolation of lignin for polycatecholamines to remove Cr (VI). J. Water Process Eng. **50**, 103334 (2022). https://doi.org/10.1016/j.jwpe.2022.103334
33. Lerma, T.A., Combatt, E.M., Palencia, M.: Novel multifunctional geomimetic soil conditioner based on multilayer hybrid composites of clay-paa-lignin: synthesis and functional characterization. Eur. Polym. J. **198**, 112376 (2023). https://doi.org/10.1016/j.eurpolymj.2023.112376
34. Wu, M., Wu, M., Pan, M., Jiang, F., Hui, B., Zhou, L.: Synthesization and characterization of lignin-graft-poly (lauryl methacrylate) via ARGET ATRP. Int. J. Biol. Macromol. **207**, 522–530 (2022). https://doi.org/10.1016/j.ijbiomac.2022.02.169
35. Huang, W., Wu, M., Liu, W., Hua, Z., Wang, Z., Zhou, L.: Value-adding of organosolv lignin: designing mechanically robust UV-resistant polymeric glass via ARGET ATRP. Appl. Surf. Sci. **475**, 302–311 (2019). https://doi.org/10.1016/j.apsusc.2018.12.266
36. Gao, S., Cheng, Z., Zhou, X., Liu, Y., Wang, J., Wang, C., Chu, F., Xu, F., Zhang, D.: Fabrication of lignin based renewable dynamic networks and its applications as self-healing, antifungal and conductive adhesives. Chem. Eng. J. **394**, 124896 (2020). https://doi.org/10.1016/j.cej.2020.124896
37. Xu, Y., Li, N., Wang, G., Wang, C., Chu, F.: Synthesis of lignin-based MMA-co-BA hybrid resins from cornstalk residue via RAFT miniemulsion polymerization and their characteristics. Polymers **13**, 968 (2021). https://doi.org/10.3390/polym13060968
38. Gebhardt, A.: Understanding additive manufacturing: Rapid Prototyping, Rapid Manufacturing, and Rapid Tooling. Hanser publications Munich, Germany (2012)
39. Horn, T.J., Harrysson, O.L.A.: Overview of current additive manufacturing technologies and selected applications **95**, 255–282 (2012). https://doi.org/10.3184/003685012X13420984463047

40. Mironov, V., Visconti, R.P., Kasyanov, V., Forgacs, G., Drake, C.J., Markwald, R.R.: Organ printing: tissue spheroids as building blocks. Biomaterials **30**, 2164–2174 (2009). https://doi.org/10.1016/J.BIOMATERIALS.2008.12.084
41. Coffigniez, M., Gremillard, L., Balvay, S., Lachambre, J., Adrien, J., Boulnat, X.: Direct-ink writing of strong and biocompatible titanium scaffolds with bimodal interconnected porosity. Addit. Manuf. **39**, 101859 (2021). https://doi.org/10.1016/j.addma.2021.101859
42. Nguyen, N.A., Barnes, S.H., Bowland, C.C., Meek, K.M., Littrell, K.C., Keum, J.K., Naskar, A.K.: A path for lignin valorization via additive manufacturing of high-performance sustainable composites with enhanced 3D printability. Sci. Adv. **4**, eaat4967 (2018). https://doi.org/10.1126/sciadv.aat4967
43. Vaidya, A.A., Collet, C., Gaugler, M., Lloyd-Jones, G.: Integrating softwood biorefinery lignin into polyhydroxybutyrate composites and application in 3D printing. Mater. Today Commun. **19**, 286–296 (2019). https://doi.org/10.1016/j.mtcomm.2019.02.008
44. Akato, K., Tran, C.D., Chen, J., Naskar, A.K.: Poly(ethylene oxide)-assisted macromolecular self-assembly of lignin in ABS matrix for sustainable composite applications. ACS Sustain. Chem. Eng. **3**, 3070–3076 (2015). https://doi.org/10.1021/acssuschemeng.5b00509
45. Sameni, J., Jaffer, S.A., Sain, M.: Thermal and mechanical properties of soda lignin/HDPE blends. Compos. Part Appl. Sci. Manuf. **115**, 104–111 (2018). https://doi.org/10.1016/j.compositesa.2018.09.016
46. Izaguirre, N., Gordobil, O., Robles, E., Labidi, J.: Enhancement of UV absorbance and mechanical properties of chitosan films by the incorporation of solvolytically fractionated lignins. Int. J. Biol. Macromol. **155**, 447–455 (2020). https://doi.org/10.1016/j.ijbiomac.2020.03.162
47. Kai, D., Chong, H.M., Chow, L.P., Jiang, L., Lin, Q., Zhang, K., Zhang, H., Zhang, Z., Loh, X.J.: Strong and biocompatible lignin /poly (3-hydroxybutyrate) composite nanofibers. Compos. Sci. Technol. **158**, 26–33 (2018). https://doi.org/10.1016/j.compscitech.2018.01.046
48. Chen, J., Qi, J., He, J., Yan, Y., Jiang, F., Wang, Z., Zhang, Y.: Biobased composites with high lignin content and excellent mechanical properties toward the ingenious photoresponsive actuator. ACS Appl. Mater. Interfaces. https://doi.org/10.1021/acsami.2c02195
49. Zhou, X., Ren, Z., Sun, H., Bi, H., Gu, T., Xu, M.: 3D printing with high content of lignin enabled by introducing polyurethane. Int. J. Biol. Macromol. **221**, 1209–1217 (2022). https://doi.org/10.1016/j.ijbiomac.2022.09.076
50. Dong, M., Li, Q., Liu, H., Liu, C., Wujcik, E.K., Shao, Q., Ding, T., Mai, X., Shen, C., Guo, Z.: Thermoplastic polyurethane-carbon black nanocomposite coating: fabrication and solid particle erosion resistance. Polymer **158**, 381–390 (2018). https://doi.org/10.1016/j.polymer.2018.11.003
51. de Leon, A.C., Chen, Q., Palaganas, N.B., Palaganas, J.O., Manapat, J., Advincula, R.C.: High performance polymer nanocomposites for additive manufacturing applications. React. Funct. Polym. **103**, 141–155 (2016). https://doi.org/10.1016/j.reactfunctpolym.2016.04.010
52. Bates, S.R.G., Farrow, I.R., Trask, R.S.: 3D printed polyurethane honeycombs for repeated tailored energy absorption. Mater. Des. **112**, 172–183 (2016). https://doi.org/10.1016/j.matdes.2016.08.062
53. Chen, Q., Mangadlao, J.D., Wallat, J., De Leon, A., Pokorski, J.K., Advincula, R.C.: 3D printing biocompatible polyurethane/poly(lactic acid)/graphene oxide nanocomposites: anisotropic properties. ACS Appl. Mater. Interfaces **9**, 4015–4023 (2017). https://doi.org/10.1021/acsami.6b11793
54. Shi, Y.-D., Cheng, Y.-H., Chen, Y.-F., Zhang, K., Zeng, J.-B., Wang, M.: Morphology, rheological and crystallization behavior in thermoplastic polyurethane toughed poly(l-lactide) with stereocomplex crystallites. Polym. Test. **62**, 1–12 (2017). https://doi.org/10.1016/j.polymertesting.2017.06.013
55. Voet, V.S.D., Guit, J., Loos, K.: Sustainable photopolymers in 3D printing: a review on biobased, biodegradable, and recyclable alternatives. Macromol. Rapid Commun. **42**, 2000475 (2021). https://doi.org/10.1002/marc.202000475
56. Sutton, J.T., Rajan, K., Harper, D.P., Chmely, S.C.: Lignin-containing photoactive resins for 3D Printing by stereolithography. ACS Appl. Mater. Interfaces **10**, 36456–36463 (2018). https://doi.org/10.1021/acsami.8b13031

57. Bassett, A.W., Honnig, A.E., Breyta, C.M., Dunn, I.C., La Scala, J.J., Stanzione, J.F.I.: Vanillin-based resin for additive manufacturing. ACS Sustain. Chem. Eng. **8**, 5626–5635 (2020). https://doi.org/10.1021/acssuschemeng.0c00159
58. Zhang, S., Li, M., Hao, N., Ragauskas, A.J.: Stereolithography 3D printing of lignin-reinforced composites with enhanced mechanical properties. ACS Omega **4**, 20197–20204 (2019). https://doi.org/10.1021/acsomega.9b02455
59. Saadi, M.A.S.R., Maguire, A., Pottackal, N.T., Thakur, M.S.H., Ikram, M.M., Hart, A.J., Ajayan, P.M., Rahman, M.M.: Direct ink writing: a 3D printing technology for diverse materials. Adv. Mater. **34**, 2108855 (2022). https://doi.org/10.1002/adma.202108855
60. Adithyamol, V.V., Gupta, S., Kandasdubramanian, B.: Revolutionizing 3D printing: harnessing the power of lignin-based ink for versatile applications. Biomed. Mater. Devices (2024). https://doi.org/10.1007/s44174-024-00193-1
61. Zhang, X., Morits, M., Jonkergouw, C., Ora, A., Valle-Delgado, J.J., Farooq, M., Ajdary, R., Huan, S., Linder, M., Rojas, O., Sipponen, M.H., Österberg, M.: Three-dimensional printed cell culture model based on spherical colloidal lignin particles and cellulose nanofibril-alginate hydrogel. Biomacromol **21**, 1875–1885 (2020). https://doi.org/10.1021/acs.biomac.9b01745
62. Roman, J., Neri, W., Fierro, V., Celzard, A., Bentaleb, A., Ly, I., Zhong, J., Derré, A., Poulin, P.: Lignin-graphene oxide inks for 3D printing of graphitic materials with tunable density. Nano Today **33**, 100881 (2020). https://doi.org/10.1016/j.nantod.2020.100881
63. Wimmers, G.: Wood: a construction material for tall buildings. Nat. Rev. Mater. **2**, 17051 (2017). https://doi.org/10.1038/natrevmats.2017.51
64. Hon, D.N.S., Shiraishi, N.: Wood and Cellulosic Chemistry, Revised, and Expanded. CRC Press (2000)
65. Thakur, M.S.H., Shi, C., Kearney, L.T., Saadi, M.A.S.R., Meyer, M.D., Naskar, A.K., Ajayan, P.M., Rahman, M.M.: Three-dimensional printing of wood. Sci. Adv. **10**, eadk3250 (2024). https://doi.org/10.1126/sciadv.adk3250
66. Liebrand, T.: 3D printed fiber reinforced lignin: exploring the options to use wood in an additive manufacturing process (2018)
67. Bierach, C., Coelho, A.A., Turrin, M., Asut, S., Knaack, U.: Wood-based 3D printing: potential and limitation to 3D print building elements with cellulose and lignin. Archit. Struct. Constr. **3**, 157–170 (2023). https://doi.org/10.1007/s44150-023-00088-7
68. Thakur, M.S.H.: Additive manufacturing of bio-inspired sustainable composites (2024)
69. Mahajan, J.S., Gottlieb, E.R., Kim, J.M., Epps, T.H.: Toward sustainable materials: from lignocellulosic biomass to high-performance polymers. Acc. Mater. Res. Accountsmr 4c00359 (2025). https://doi.org/10.1021/accountsmr.4c00359
70. Ramaux, J., Ziegler-Devin, I., Besserer, A., Nouvel, C.: 3D printing of wood composites: state of the art and opportunities. Polymers **16**, 2827 (2024). https://doi.org/10.3390/polym16192827
71. Farooq, M., Zou, T., Riviere, G., Sipponen, M.H., Österberg, M.: Strong, ductile, and waterproof cellulose nanofibril composite films with colloidal lignin particles. Biomacromol **20**, 693–704 (2019). https://doi.org/10.1021/acs.biomac.8b01364
72. Kai, D., Tan, M.J., Chee, P.L., Chua, Y.K., Yap, Y.L., Loh, X.J.: Towards lignin-based functional materials in a sustainable world. Green Chem. **18**, 1175–1200 (2016). https://doi.org/10.1039/C5GC02616D
73. Gaspar, R., Fardim, P.: Lignin-based materials for emerging advanced applications. Curr. Opin. Green Sustain. Chem. 41100834 (2023). https://doi.org/10.1016/j.cogsc.2023.100834

Dr. Md Shajedul Hoque Thakur is a Postdoctoral Fellow in Mechanical and Aerospace Engineering at University of Houston under the supervision of Dr. Rahman. His research focuses on AI-driven advanced manufacturing of sustainable materials. Dr. Thakur received his Ph.D. in Materials Science and NanoEngineering from Rice University in 2024, where he invented the first-ever 3D printing of wood structures using water-based inks. Dr. Thakur received his B.Sc.

in Mechanical Engineering from Bangladesh University of Engineering and Technology in 2021, where he worked on nanoscale thermal transport, mechanics of functionally graded nanowires, and deep learning for accelerating molecular dynamics simulations.

Md Tofazzal Hossain is a Ph.D. student in Mechanical and Aerospace Engineering at the University of Houston, supervised by Dr. Rahman, whose research centers on advanced manufacturing of composite and biodegradable materials. He concurrently serves as an assistant professor in Mechanical Engineering at Ahsanullah University of Science and Technology, Bangladesh. Hossain earned his M.Sc. (2022) and B.Sc. (2017) in Mechanical Engineering from Bangladesh University of Engineering and Technology. He investigated finite-element models of the human left ventricle for post-AMI heart-failure prognosis and CFD-based design of naturally ventilated residential buildings. Committed to sustainability, he views machine learning as a transformative driver for next-generation manufacturing and materials science.

Mir Muhammad Fahad is a Ph.D. student in the Department of Mechanical and Aerospace Engineering at University of Houston, supervised by Dr. Rahman. He received his B.Sc. in Textile Engineering from Bangladesh University of Textiles, Bangladesh in 2018 and his M.S. in Advanced Materials Engineering for Information and Electronics from Kyung Hee University, South Korea in 2022. Fahad specializes in the fundamental studies of polymer science and its applications in textile chemistry, hybrid composites, biomedical sensors/ energy harvesting devices, and thermodynamics driven cryopreservation. He is currently developing protein-based nanocomposite films for preserving agro-products under Dr. Rahman's supervision.

Dr. Muhammad M. Rahman is an Assistant Professor in the Mechanical and Aerospace Engineering Department at the University of Houston, Texas. He received his Ph.D. from Cornell University, where he studied the structure-property relation-ship in biomaterials and nanocomposites. Dr. Rahman has significant experiences in manufacturing and characterizing nanocomposites. He designs next-generation nanocomposites via advanced manufacturing technologies, including additive manufacturing, emphasizing sustainability for load-bearing structures, and environmental applications. He has published articles in many prestigious journals, including Nature Communications, Science Advances, Advanced Materials, ACS Nano, etc. His research achievements have been featured by Science, Nature, BBC, New Scientist, and many other media outlets. Dr. Rahman has received several federal and industrial grants from the USA and Canada, as a Principal Investigator.

Part III
Applications

Chapter 8
Three-Dimensional Printing of Biodegradable Materials in Biomedical Sector

Karthik Tappa

Abstract The utilization of 3D printing technology for fabricating biodegradable materials has revolutionized medical applications, offering unprecedented flexibility and customization in healthcare. Biodegradable materials are gaining traction due to their ability to degrade harmlessly within the body, reducing the need for surgical interventions to remove implants. 3D printing enables the precise fabrication of complex geometries tailored to patient-specific needs, promoting personalized medicine. In orthopedics, for instance, biodegradable implants can mimic the mechanical properties of bone and gradually degrade as new tissue forms, supporting natural healing processes. In addition to implant customization, 3D printed biodegradable materials play a pivotal role in drug delivery systems. They facilitate the controlled release of therapeutic agents, ensuring precise dosing and minimizing systemic side effects. Such capabilities are crucial in personalized medicine, where patient-specific treatments can be developed based on individual physiological and pathological characteristics. Beyond implants and drug delivery, 3D printing of biodegradable materials holds promise in tissue engineering by creating scaffolds that support cellular growth and tissue regeneration. These scaffolds can be designed with intricate internal architectures to mimic natural tissues, facilitating integration and functionality. Challenges including material selection, degradation kinetics, and regulatory considerations remain, yet ongoing research and technological advancements continue to address these hurdles. As such, the convergence of 3D printing and biodegradable materials represents a transformative approach in advancing personalized medicine, improving patient outcomes, paving the way for more effective therapies, and shaping the future of healthcare.

Keywords 3D printing · Patient-specific implants · Tissue engineering · Drug delivery

---

K. Tappa (✉)
Department of Breast Imaging, Division of Diagnostic Imaging, The University of Texas MD Anderson Cancer Center, Houston, TX, USA
e-mail: kktappa@mdanderson.org

© The Author(s), under exclusive license to Springer Nature Singapore Pte Ltd. 2026

C. Klemmt and G. Beaucage (eds.), *Biodegradable 3D-Printing*, The Materials Research Society Series, https://doi.org/10.1007/978-981-95-4025-9_8

1 Introduction

The advent of three-dimensional (3D) printing has revolutionized the biomedical field, offering unparalleled opportunities to address complex healthcare challenges through innovative fabrication techniques. Among these, the integration of biodegradable materials with 3D printing technology has emerged as a groundbreaking approach, redefining the possibilities for personalized medicine. Biodegradable materials, characterized by their ability to safely decompose within biological systems, present significant advantages in medical applications, ranging from implants to tissue engineering. These materials eliminate the need for secondary surgical interventions for removal, thereby reducing patient risk and improving recovery outcomes [1].

The versatility of 3D printing enables the precise construction of intricate geometries tailored to patient-specific requirements. This capability supports the creation of highly customized solutions, such as orthopedic implants that mimic the mechanical properties of bone or scaffolds that facilitate tissue regeneration [2]. Furthermore, 3D printed biodegradable materials hold immense potential in controlled drug delivery systems, ensuring precise therapeutic dosing and minimizing adverse systemic effects [3].

Despite these promising developments, challenges persist in material selection, degradation kinetics, and regulatory compliance, which must be addressed to ensure the widespread adoption of these technologies [4]. Through this chapter, we explore the confluence of 3D printing and biodegradable materials, examining their transformative role in personalized medicine, their applications across various medical domains, and the ongoing efforts to overcome existing hurdles. This chapter sets the stage for a comprehensive analysis of the technological advancements, clinical implications, and future prospects that this innovative synergy offers in shaping the future of healthcare.

1.1 Overview of 3D Printing in Biomedical Applications

3D printing, also known as additive manufacturing, has gained significant traction in the biomedical sector due to its ability to fabricate complex structures with precision and customization. This technology employs layer-by-layer construction to create intricate designs from digital models, enabling the production of patient-specific solutions. Its application spans various medical domains, including surgical guides, prosthetic, orthotics, drug delivery systems, anatomical models, tissue engineering, and regenerative medicine [5]. Figure 1 details some of the applications of 3D printing in medicine.

In orthopedics, 3D printing facilitates the creation of personalized implants and surgical guides, optimizing the fit and function for individual patients [6]. In prosthetics, it enables the fabrication of affordable and customized prosthetic limbs

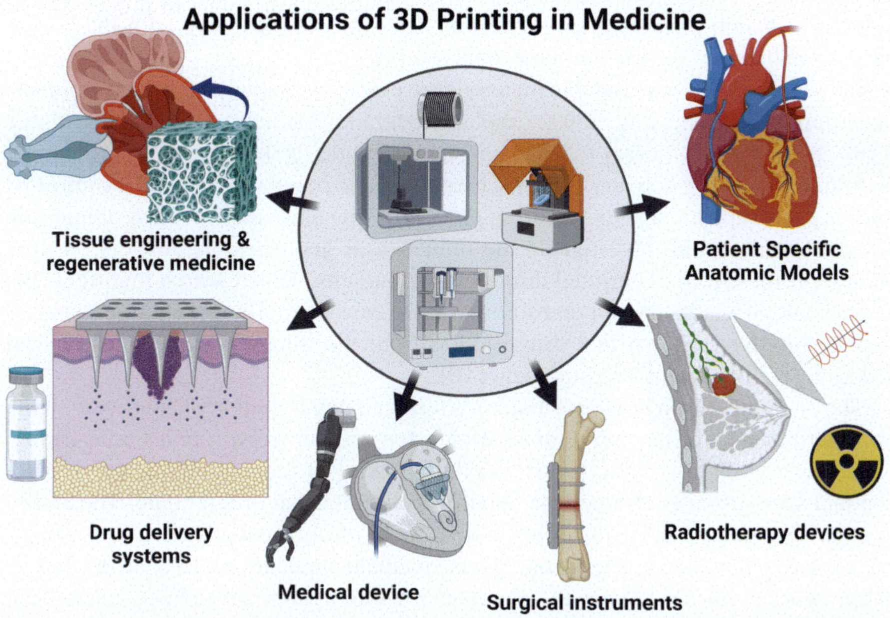

Fig. 1 Applications of 3D printing in medicine. Created in BioRender. Tappa (2025) https://BioRender.com/14y9shu

tailored to the user's anatomy and functional needs [7]. Dentistry has also leveraged 3D printing for producing dental implants, crowns, and aligners with exceptional accuracy [8]. Moreover, 3D printing plays a pivotal role in tissue engineering by generating scaffolds that replicate the architecture of natural tissues, supporting cellular growth and regeneration [2].

The ability of 3D printing to integrate multiple materials and functions within a single structure makes it a versatile tool for biomedical innovation. Additionally, its digital workflow allows for rapid prototyping, accelerating research, and development cycles while reducing costs. These advantages position 3D printing as a transformative technology in advancing medical science and improving patient care.

1.2 Importance of Biodegradable Materials in Medicine

In conjugation with 3D printing technology, biodegradable materials have emerged as a cornerstone of innovation in modern medicine due to their unique ability to degrade safely and harmlessly within the human body. These materials eliminate the need for secondary surgical procedures to remove temporary implants, reducing patient trauma, surgical risks, and healthcare costs. Their biodegradation can be tailored to match physiological processes, ensuring compatibility with the body's natural healing and regeneration mechanisms [9].

In orthopedics, biodegradable materials are extensively used to fabricate implants that support bone healing. These implants degrade gradually as new bone tissue forms, providing temporary mechanical support while facilitating natural recovery.

Similarly, in drug delivery systems, biodegradable polymers enable the controlled release of therapeutic agents, ensuring precise dosing and minimizing systemic side effects. Domsta et al. investigated the influence of geometric parameters on drug release profiles from 3D printed drug-eluting implants. This research highlights the critical role of geometry in controlling drug release from 3D printed implants and their results have shown that drug elution from the implants can be customized according to the patient's requirement [10].

The biocompatibility of these materials further extends their applicability to tissue engineering, where they serve as scaffolds for cellular growth and tissue formation. These scaffolds mimic the extracellular matrix of natural tissues, providing a conducive environment for cells to adhere, proliferate, and differentiate. The ability to engineer biodegradable materials with specific mechanical and chemical properties enhances their versatility across diverse medical applications. Li et al. developed 3D printed porous Ti6Al4V scaffolds coated with polydopamine (PDA) to enhance their osteogenic properties. The 5-week in vivo rabbit study, comparing these 3D printed coated implants and non-coated implants, have shown enhanced cell adhesion, stress shield reduction and early bone growth for PDA coated implant cohorts, promoting early osteogenesis [11].

The importance of biodegradable materials is amplified by numerous researchers in the context of personalized medicine, where treatments are tailored to individual patient needs. By combining biodegradability with the precision of 3D printing, medical professionals can create bespoke solutions that align with patient-specific physiological and pathological characteristics, ushering in a new era of customized and effective therapies.

2 Biodegradable Materials in the Biomedical Sector

Biodegradable materials have garnered immense attention in the biomedical field due to their ability to support natural healing and eliminate the need for invasive removal procedures. This section explores the properties of biodegradable materials, their advantages over conventional materials, and their diverse applications in modern medicine. Based on their origin, structure and functional properties, biodegradable materials can be classified into the following categories:

Natural Polymers: Natural polymers are derived from biological sources and exhibit excellent biocompatibility and biodegradability. These materials are known to mimic natural extracellular matrices, making them suitable for soft tissue engineering, wound healing, and drug delivery. Polymers including collagen, chitosan, gelatin, alginate and hyaluronic acid (HA) are some of the most common biomaterials that are used widely in 3D printing.

In a recent study conducted by Damle et al. collagen derived from extracellular matrix of chicken skin was used to 3D print artificial skin grafts. This collagen bioink when incorporated with polyvinyl alcohol (PVA) and gelatin exhibited excellent mechanical strength, significant cell proliferation and migration and accelerated wound healing properties [12].

Dinh et al. used combination of natural polymers including chitosan, pectin and sodium alginate to 3D print custom wound dressing that can elute sustained amounts of chemotherapeutic drugs for postoperative care in colorectal cancer patients treatment [13].

Using extrusion-based bioprinting techniques, Demirel et al. 3D printed biomimetic scaffolds with varying concentration of hyaluronic acid (HA) to enhance the formation of liver cancer spheroids that can serve as promising models for mimicking solid liver tumors, providing a valuable platform for studying cancer stem cell properties and aiding in development of targeted drug therapy [14].

Synthetic Polymers: Synthetic polymers are made with controlled chemical synthesis and offer excellent properties including tunable degradation rates and mechanical properties, making them versatile for numerous biomedical applications. Polylactic acid (PLA), polyglycolic acid (PGA), polycaprolactone (PCL), and poly(lactic-co-glycolic acid) (PLGA) are some of the synthetic polymers that are commonly used in 3D printing for biomedical applications.

Gregor et al. demonstrated that PLA can be used in commercially available 3D printers and can serve as viable candidates for bone tissue engineering. Their article suggests that with appropriate design considerations, PLA scaffolds have the ability to mimic mechanical properties of bone and can also support cellular activities crucial for tissue regeneration [15].

Bassand et al. loaded ibuprofen into PLGA polymer and used a droplet deposition type 3D printing technique to fabricate mesh shaped PLGA implants with varying densities. Their drug release studies demonstrated significant changes in drug elution with implant density values offering a promising approach for developing patient-specific drug delivery systems [16].

Bioceramic-Based Materials: Bioceramic materials have the tendency to mimic the mineral composition of natural bone and are often combined with other biodegradable polymers to enhance mechanical strength and osteoconductivity, particularly for bone-related applications. Calcium Silicates, calcium phosphates and hydroxyapatite are the most common bio-ceramics used for 3D printing in bone tissue regeneration.

Using extrusion-based 3D printing techniques, Wattanaanek et al. fabricated multilayer scaffolds composed of calcium phosphate and calcium sulfate materials integrated with alginate and cellulose to achieve precise control over the scaffold's architecture. Their article suggests that this combination of calcium-based minerals with alginate and cellulose provided increased mechanical properties of the scaffolds as well as enhanced osteogenic differentiation. These findings suggest that bioceramic scaffolds could serve as effective bone graft substitutes, offering a customizable and biocompatible solution for bone regeneration applications [17].

Hydrogels: Biodegradable hydrogels are water-absorbing networks of polymers that provide a hydrated environment suitable for cell growth and drug delivery. Gelatin Methacrylate (GelMA), a photo-crosslinkable material, is most commonly used in 3D printing, either stand-alone or in combination with polyethylene glycols [18] or chitosan-alginate hydrogels [19].

Using 3D printing, Sun et al. designed a bilayer hydrogel scaffold with GelMA integrated with black phosphorus (BP), β-tricalcium phosphate (β-TCP), and exosomes that mimics the natural cartilage-bone interface supporting both cartilage and bone regeneration. There in vitro studies demonstrated proliferation and differentiation of chondrocytes and osteoblasts, whereas their in vivo studies showed simultaneous regeneration of cartilage and subchondral bone, leading to improved integration at the cartilage-bone interface [20].

Composite Materials: Biodegradable composites are hybrid materials that combine different biodegradable materials to leverage their individual advantages, such as enhanced mechanical strength or bioactivity. PLA/PCL blends [21], HA/PLGA composites [22], and chitosan/collagen mixtures [23] are the most common 3D printing composite materials used for numerous biomedical applications.

Cresswell-Boyes et al. employed a multi-material 3D printing technique using a digital light processing (DLP) 3D printer to create artificial teeth with a composite of materials including hydroxyapatite, bio glass, and ceramic materials. This study has shown feasibility of using composite 3D printing to fabricate artificial teeth that closely mimic the structural and mechanical properties of natural human teeth [24].

Using FDM-based 3D printing techniques, we explored the feasibility of creating personalized, biodegradable implants capable of delivering estrogen and/or progesterone for obstetric and gynecologic applications. The author's team utilized polycaprolactone (PCL), a biodegradable polymer, as the base material, incorporating hormones such as estrone (E1), estradiol (E2), estriol (E3), and progesterone (P) into the PCL matrix. These hormone-infused PCL filaments were extruded and then 3D printed into various medical device shapes, including surgical meshes, subdermal rods, intrauterine devices (IUDs), and pessaries. To assess the performance of these constructs, hormone-release studies using enzyme-linked immunosorbent assays (ELISA) over a one-week period were conducted, demonstrating sustained hormone elution. Thermal properties were evaluated using thermogravimetric analysis and differential scanning calorimetry, confirming that the printing process temperatures did not degrade the hormones. Biocompatibility and bioactivity were assessed through luciferase assays, which showed increased activity without adverse effects, indicating the constructs' safety and effectiveness. The results from this study concluded that 3D printing hormone-eluting, biodegradable implants is a viable approach, offering the potential for personalized medicine solutions in obstetrics and gynecology. These custom-designed devices could provide tailored hormone delivery and fit, enhancing patient-specific treatment options [25].

2.1 Properties and Advantages of Biodegradable Materials

Biodegradable materials are specifically designed to decompose into non-toxic byproducts upon interaction with biological systems. This process, often driven by enzymatic reactions or hydrolysis, ensures that the material breaks down into harmless substances such as carbon dioxide, water, or naturally occurring metabolic intermediates.

Key properties of biodegradable materials include:

Biocompatibility: They interact safely with bodily tissues and fluids without inducing adverse immune responses [26]. This biocompatibility ensures that these materials do not cause inflammation, toxicity, or rejection when implanted or used in medical applications. Lactic acid-based polymers including PLA, PCL, and PLGA, are the most commonly used biodegradable material in medical 3D printing due to their ability to degrade in situ by a hydrolysis process producing easily metabolizable byproducts including lactic acid, water, and carbon dioxide [5].

Tailorable Degradation Rates: By modifying the composition and structure of biodegradable materials, their degradation rates can be adjusted to match the body's natural healing timeline. For example, PLGA is widely researched due to its biocompatibility nature. This polymer's degradation rate can be adjusted by modifying its molecular weight and the ratio of lactic to glycolic acid, allowing for a precise control over drug release timelines [5].

Mechanical Versatility: These materials can be engineered to mimic the structural and mechanical properties of tissues such as bone, cartilage, or skin. For instance, composite materials combining polymers with bio-ceramics, including HA, are commonly used for bone tissue engineering to mimic the osteoconductive and mechanical properties of bone [27].

The advantages of biodegradable materials in medical applications include:

Elimination of Secondary Surgeries: Since the material degrades naturally, patients are spared additional surgeries to remove temporary implants. In a case study conducted by Fidanza et al. on a severe acetabular bone loss case, a custom-made, 3D printed implant allowed for successful reconstruction showing that in prosthetic revision surgeries with significant bone loss, a 3D-printed implant offers a viable solution exhibiting excellent outcomes [28].

Reduced Risk of Chronic Inflammation: Biodegradable materials are designed to degrade naturally within the body into non-toxic by-products, such as water and carbon dioxide, that can be safely metabolized or excreted. This property significantly minimizes the long-term exposure of foreign substances to the immune system, thereby reducing the risk of chronic inflammation. Unlike permanent implants, which may cause irritation, fibrosis, or an ongoing immune response, biodegradable implants dissolve in harmony with the healing process. For instance, PLGA scaffolds used in tissue engineering degrade over a defined period, providing mechanical support only during critical healing phases. Its degradation occurs via hydrolysis of its ester bonds, producing the monomers lactic acid and glycolic acid. These by-products seamlessly integrate into the body's natural metabolic pathways, specifically the Krebs cycle, where they are further broken down into water (H_2O) and carbon dioxide (CO_2), both of which are non-toxic and readily excreted [29]. This self-resolving behavior ensures that the body does not experience prolonged exposure to synthetic materials, reducing the chances of implant-related complications such as chronic inflammation or infection.

Cost Efficiency: The use of biodegradable materials in medical applications not only improves patient outcomes but also contributes to significant cost savings in healthcare [30]. Traditional implants often necessitate secondary surgeries for removal once their function is complete, adding to the overall treatment cost and patient recovery time. In a retrospective study conducted by Nudelman et al. on pediatric patients with distal tibial fractures, one cohort treated with traditional titanium or steel screws and another cohort treated with PLGA implants have shown that PLGA implants did not interfere with the growth plate and eliminated the need for secondary surgeries to remove the hardware. This approach potentially reduced patients' burden, lowering healthcare costs, and decreasing risks associated with additional procedures, including infections or complications from general anesthesia [31]. Additionally, advancements in scalable manufacturing techniques for biodegradable materials have further decreased production costs, making them accessible for broader clinical applications.

By leveraging their unique properties and advantages, biodegradable materials are reshaping the biomedical field, driving innovation in personalized medicine, and addressing the growing demand for sustainable healthcare solutions.

3 3D Printing Technology for Biodegradable Materials

Biodegradable materials represent a transformative shift in the biomedical landscape, offering innovative solutions that are both functional and sustainable. Their ability to degrade naturally within biological environments without causing harm has propelled their application across diverse medical fields. This section delves into the properties, advantages, and multifaceted applications of biodegradable materials in medicine. The integration of 3D printing technologies with biodegradable materials marks a significant advancement in biomedical engineering. This combination leverages the precision and versatility of 3D printing to create highly customized and functional solutions tailored to individual patient needs. This section explores the techniques, methods, and advantages of using 3D printing for biodegradable materials.

3.1 Techniques and Methods in 3D Printing

3D printing technology encompasses a variety of techniques, each suited for specific applications and material types. The following methods are commonly employed in the fabrication of biodegradable materials:

Fused Deposition Modeling (FDM): FDM is a widely used additive manufacturing technique that extrudes thermoplastic filaments layer-by-layer to create a 3D object. Filaments made from polylactic acid (PLA), polycaprolactone (PCL), and other biopolymers are commonly utilized. A biodegradable filament is fed into a heated nozzle where it is melted. The molten material is deposited onto a build platform following a pre-programmed path. This process repeats layer-by-layer until the structure is completed. This type of technology is widely accepted due to their compatibility with numerous biodegradable polymers and affordability. But this technology suffers limitations, including a limited resolution compared to other 3DP technologies, and time-consuming post-processing. In clinical settings, FDM technologies are commonly used for fabricating porous structures that mimic trabecular bone for tissue engineering [32–34] and localized drug release applications [35, 36]. Figure 2 illustrates the print head of a FDM 3D printer.

Stereolithography (SLA): SLA employs a laser to cure liquid photosensitive resins layer-by-layer into solid structures. In this technology, the build platform is usually submerged in a vat of liquid resin and a laser selectively cures specific areas of

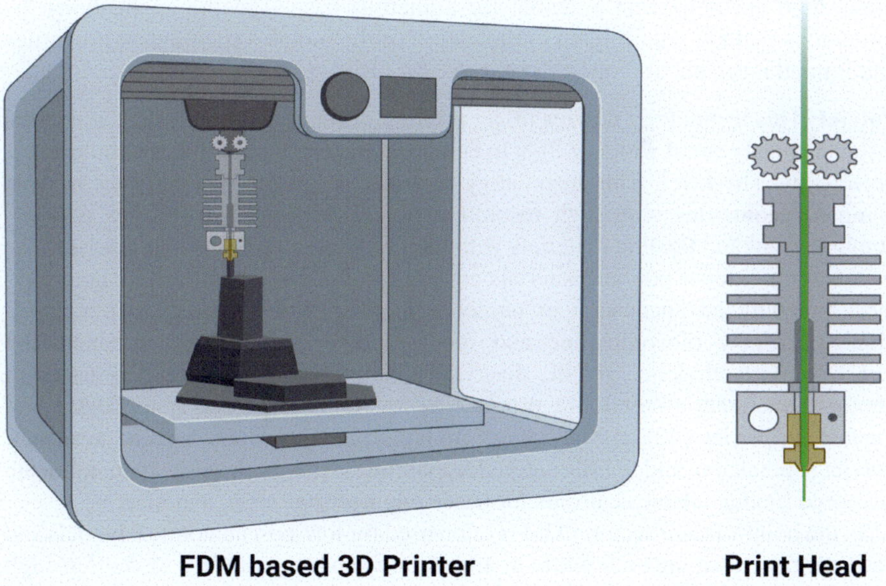

Fig. 2 Illustration of FDM-based 3D printer and its print head showing extrusion of 3D printer filament

the resin, solidifying it layer-by-layer until the whole object is formed. Due to this technique, high resolution 3D printing is possible and makes SLA ideal for intricate designs and applications requiring smooth surfaces [37]. Today, in the medical sector, SLA is widely used for creating dental restorations and prosthetics [38]. Although this technology has numerous advantages, it suffers from having limited availability of photosensitive resin varieties.

Selective Laser Sintering (SLS): SLS employs a laser to sinter powdered biodegradable materials, allowing for the creation of complex geometries without the need for support structures. In this technology, a thin layer of powdered biodegradable material, such as polyhydroxyalkanoates, is spread over the build platform and a laser selectively fuses the powder particles, in accordance with the digital tool path assigned by the CAM software, while lowering the build platform incrementally. This process repeats until the object is fully formed. This technology is currently used for the fabrication of high-strength, bioresorbable implants, and patient-specific prosthetic devices [39].

Extrusion-Based Bioprinting: This method prints bioinks composed of biodegradable polymers and living cells, facilitating the creation of tissue-engineered constructs, as shown in Fig. 3. In this process, a bioink is extruded through a nozzle and deposited onto the build platform layer-by-layer. Simultaneous crosslinking (via physical or chemical methods) of this construct is done in order to maintain its shape and mechanical integrity. These bioprinted constructs are cultured under controlled environment to encourage cell growth and tissue formation. This technique is most commonly used to engineer scaffolds for tissue regeneration [40]. This technique offers high customization potential for numerous tissue-specific applications. Its limitations include requiring an optimization of the bioink's rheological properties and limited mechanical properties of the scaffolds [41].

Polyjet: This technology utilizes inkjet printing of photopolymer resins, which are instantaneously cured with UV light to build precise, multi-material, and multi-color objects layer-by-layer. This technology is widely used for its ability to produce complex geometries with high resolution and smooth surface finishes and can combine rigid and flexible materials with various shore values in a single print.

While traditional polyjet materials are primarily non-biodegradable photopolymers, ongoing advancements in photopolymerization are leading to the development of new biocompatible and biodegradable resins. Materials including PEGDA (polyethylene glycol diacrylate) have been successfully used in photopolymerization-based 3D printing to create biodegradable scaffolds with tunable properties [42]. As new biodegradable photopolymers become available, polyjet's precision and multi-material capabilities can be leveraged to fabricate advanced biodegradable scaffolds for tissue engineering.

In the point-of-care sector, FDM is often used for low-cost anatomical models and basic surgical planning tools due to its affordability and accessibility. SLA offers high precision and smooth surface finishes, making it suitable for dental models, surgical guides, and transparent anatomical parts. Polyjet provides the highest resolution

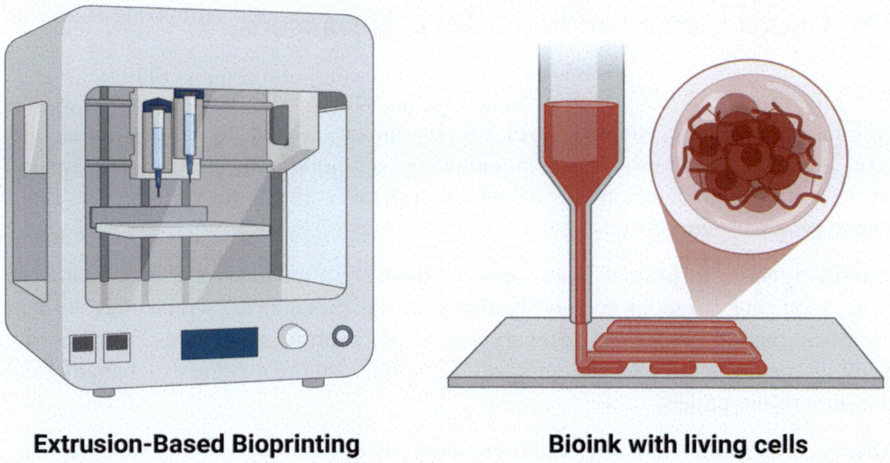

Fig. 3 Illustration of an extrusion-based bioprinter and extrusion nozzle filled with biodegradable polymer incorporated with living cells (Bioink)

and multi-material capabilities, ideal for realistic, multi-textured anatomical models and complex pre-surgical simulations, though it is more expensive and maintenance intensive. Figure 4 shows images of anatomical models 3D printed on FDM, SLA, and Polyjet 3D printers.

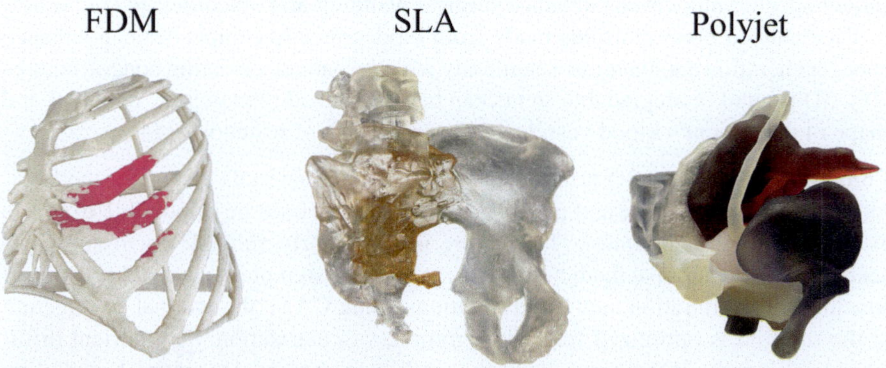

Fig. 4 Images of anatomical models printed on various 3D printing technologies. The FDM 3D printed model shows ribcage in white and tumor on the ribs in magenta color. The SLA 3D printed model of the hemipelvis shows healthy bone in clear and tumor in sacrum in color. The polyjet 3D printed model shows red color tumor present in between bladder and rectum. The pelvic floor is in white, sacrum in clear, ureters in translucent, and bladder in light pink

3.2 Customization and Flexibility in Healthcare

The integration of 3D printing technologies with biodegradable materials has introduced an unprecedented level of customization and flexibility in healthcare. In contrast to traditional manufacturing techniques, 3D printing enables the creation of patient-specific medical devices, implants with complex geometries, rapid prototyping and reduced cost:

Patient-Specific Implants: Traditional implants are often designed in standardized sizes, requiring surgeons to modify them during procedures, which may lead to complications and extended recovery times. 3D printing eliminates this issue by enabling the production of patient-specific implants that match the exact anatomical structure of the patient.

Advanced imaging techniques including computed tomography (CT) and magnetic resonance imaging (MRI) are used to develop detailed 3D models using segmentation software. These generated 3D models are processed for designing matching implants using computer-aided design (CAD) software. These custom implants are then 3D printed, sterilized and directly implanted reducing intraoperative modifications.

Biodegradable 3D printed implants can be customized to fit complex facial and skull defects after trauma and surgery. The research study conducted by Bergeron et al. on patients requiring immediate surgical intervention for cranio-maxillofacial (CMF) trauma demonstrated that in-house 3D printing is a viable and efficient approach for acute CMF trauma surgeries. The ability to rapidly produce patient-specific models can enhance surgical planning and outcomes. [43]

Personalized bone scaffolds made from biodegradable composites can enhance bone regeneration and integrate seamlessly with the patient's existing bone structures [28]. 3D printed biodegradable stents can be customized to match the diameter and shape of the patient's blood vessels, reducing the risk of restenosis [44].

Complex Geometries: 3D printing of biodegradable materials allows for the fabrication of highly complex and porous structures that would be impossible to create using traditional techniques. This feature is particularly valuable for tissue engineering and regenerative medicine, where scaffold design plays a crucial role in cell attachment, proliferation, and differentiation. Using CAM, the internal architecture of the scaffold is optimized for cell migration, vascularization, and nutrient diffusion. Additionally, 3D printing allows the scaffold to be printed in materials including PLGA, collagen, hydroxyapatite, or bioinks that can support cellular growth. Raja et al. investigated the effects of varying porosity levels and phase compositions in 3D printed calcium phosphate scaffolds on bone tissue regeneration. In this study, the researchers utilized 3D printing technology to fabricate scaffolds with controlled porosity and different phase compositions of calcium phosphate bone cements. Their findings indicate that optimizing these parameters can enhance the scaffolds compressive strength without compromising porosity, leading to more efficient bone tissue regeneration [45]. The key advantages include enhancing tissue growth and vascularization by customizing pore size and structure, customizing regeneration time by tailoring scaffold's degradation rate and reducing risk of implant rejection by promoting natural tissue formation.

Rapid Prototyping: Before 3D printing, developing custom medical devices, prosthetics and implants was time-consuming and expensive. 3D printing accelerates innovation by allowing rapid prototyping and iterative testing at a fraction of the cost. The general workflow of rapid prototyping includes designing a 3D model of the prototype digitally using CAD software, 3D printing multiple iterations, prototype testing and finally optimization.

This 3D printing reduces the time required to develop personalized prosthetic limbs and orthopedic devices. Patient-specific 3D printed anatomical models help surgeons plan complex procedures more effectively and customizable 3D printed biosensors allow for real-time health monitoring.

Mills et al. employed FDM-based 3D printing technology to test the feasibility of creating customized nasal stents tailored to individual patient's anatomy. The researchers utilized polylactic acid (PLA) filament, a biodegradable and biocompatible thermoplastic, to fabricate nasal stents using fused deposition modeling (FDM) 3D printing. The authors have designed the stents based on anatomical data to ensure a proper fit within the nasal cavity. This study involved producing multiple stent prototypes and evaluating their structural integrity, flexibility, and suitability for nasal applications. The results from the stents evaluation demonstrated that the 3D printed PLA stents possessed adequate mechanical strength and flexibility, making them suitable for maintaining nasal patency post-surgery. The customization potential of 3D printing allowed to produce stents that conformed closely to patient-specific nasal anatomy, potentially improving patient comfort and surgical outcomes. The study concluded that 3D printing is a viable method for fabricating personalized nasal stents, offering advantages in customization, production speed, and cost-effectiveness. This approach could enhance postoperative care in nasal surgeries by providing tailored stenting solutions that align with individual patient needs [46].

By combining the adaptability of 3D printing with the sustainability of biodegradable materials, this technology addresses the growing demand for personalized and eco-friendly healthcare solutions. From patient-specific implants and complex tissue scaffolds to personalized devices, these advancements are driving precision medicine and regenerative therapies to new heights.

4 Applications in Orthopedics

The field of orthopedics has seen significant advancements through the integration of 3D printing and biodegradable materials. These innovations offer solutions that are tailored to individual patient needs, enhancing treatment efficacy and reducing recovery times [47]. This section explores the specific applications and benefits of biodegradable materials in orthopedics.

4.1 Biodegradable Implants

Traditional orthopedic implants, made of metals including titanium, stainless steel, and cobalt-chromium alloys, often require surgical removal after healing, leading to additional procedures, increased costs, and potential complications. In contrast, biodegradable implants made from material including PLA, PLGA, and magnesium alloys dissolve naturally in the body [5]. Biodegradable implants are revolutionizing orthopedic care by providing temporary mechanical support while enabling natural tissue regeneration. These implants are designed to degrade gradually within the body, matching the pace of bone or tissue healing [48]. Key applications include:

Bone Fixation Devices: Bone fixation devices play a critical role in stabilizing fractures and promoting bone healing. Traditionally, metallic fixation devices including titanium screws, plates and pins have been used. However, biodegradable bone fixation devices eliminate the need for secondary surgeries to remove implants after healing, reducing patient burden and healthcare costs [49].

Biodegradable screws and plates made from PLA, PLGA, or magnesium alloys, provide temporary support to fractured bones while gradually degrading over time. Bioabsorbable pins and rods are used in pediatric orthopedics, as they degrade without interfering with bone growth and maintain mechanical integrity during the initial healing phase. Bioresorbable plates for internal fixation are ideal for complex fractures in skull and mandible. These are designed to provide high mechanical strength and controlled degradation rates [50].

Joint Reconstruction: Joint reconstruction is a complex procedure involving the repair or replacement of damaged joints due to osteoarthritis, trauma, or congenital defects. Traditional joint implants are metallic, but biodegradable materials are now being explored to reduce implant-related complications and promote natural joint regeneration. [9]

Spinal Fusion: Spinal fusion is a surgical procedure used to stabilize the spine by fusing two or more vertebrae together. Biodegradable materials in spinal fusion implants offer a promising alternative to traditionally metallic cages by reducing the risk of implant failure and promoting bone regeneration [51].

4.2 Mimicking Mechanical Properties of Bone

One of the critical advantages of biodegradable materials in orthopedics is their ability to mimic the mechanical properties of natural bone. This ensures that implants provide sufficient support without causing stress shielding, where an implant carries too much load, preventing proper bone growth. By matching the stiffness and strength of bone, these materials promote natural healing and reduce the risk of implant failure. Material selection, structural optimization and use of appropriate printing technology are some of the key considerations that must be done in order to achieve maximum mimicking of the scaffolds to the native tissue. In a research study conducted by Yang et al., a novel hybrid bioactive scaffold that mimics the natural bone architecture has been developed utilizing biopolymers including pearl powder, fish gelatin methacrylate (GelMa) and vascular endothelial growth factor (VEGF) in microfluidic-assisted 3D printing technology. Their results indicate that the scaffolds demonstrated enhanced bioactivity, promoting cell proliferation and differentiation, crucial for bone mimicking and healing [52].

Similarly, Pan et al. developed a series of microRNA (miRNA) activated hydrogel scaffolds with varying crosslinking degrees to investigate the balance between scaffold degradation and miRNA release on osteogenesis mimicking physiological bone healing processes. They found that scaffolds with a higher crosslinking degree exhibited a more suitable degradation rate and miRNA release profile, leading to more efficient osteoinductive behavior making it the preferred scaffold for accelerated bone regeneration [53].

4.3 Supporting Natural Healing Processes

Biodegradable materials in orthopedics not only provide structural support but also actively contribute to the healing process. They can be engineered to release growth factors or other bioactive agents that stimulate bone growth and tissue repair [54–58]. This integrated approach ensures a seamless transition from implant support to natural recovery, improving patient outcomes and reducing recovery times.

5 Role in Drug Delivery Systems

Biodegradable materials have redefined the field of drug delivery systems, offering unprecedented control over the release of therapeutic agents. These materials enable the creation of patient-specific treatment plans that optimize therapeutic outcomes while minimizing side effects. This section explores their applications and benefits in drug delivery systems.

5.1 Controlled and Sustained Drug Release

One of the most significant advantages of biodegradable materials in drug delivery is their ability to provide controlled and sustained release of therapeutic agents. Key applications include:

Localized Drug Delivery: Biodegradable polymers can deliver drugs directly to the target site, reducing systemic exposure and associated side effects [59].

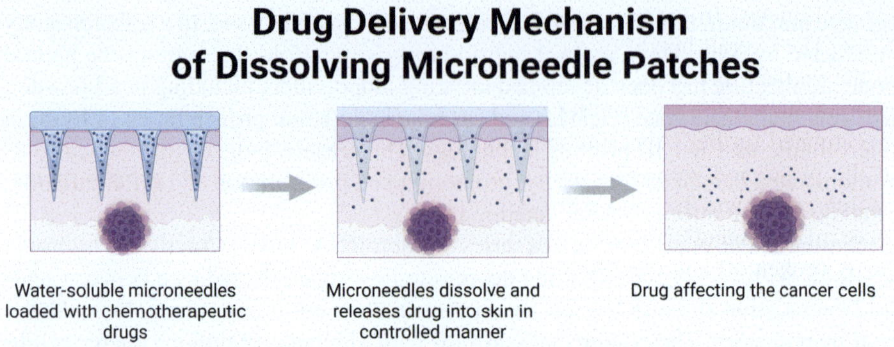

Fig. 5 Image showing steps involved in localized delivery of chemotherapeutic drugs from 3D printed dissolvable microneedle patches. Image created in BioRender. Tappa (2025)

3D printing technology has revolutionized microneedle (MN) design by enabling precise, customizable structures for biomedical use. Key factors including geometry, tip sharpness, and material choice can be customized according to the requirement ensuring effective drug delivery and skin penetration [60]. Figure 5 shows a mechanism of action of 3D printed dissolvable microneedle patches loaded with chemotherapeutic drugs.

Prolonged Therapeutic Effects: By tailoring the degradation rate of the material, it is possible to achieve a steady release of drugs over an extended period, reducing the frequency of administration [61].

Multi-Drug Delivery Systems: These materials can be engineered to deliver multiple drugs simultaneously or in a sequential manner, addressing complex medical conditions [62].

5.2 *Personalized Medicine Applications*

Biodegradable drug delivery systems play a crucial role in advancing personalized medicine by enabling treatments that are tailored to individual patients' needs. Specific applications include:

Cancer Therapy: Targeted delivery of chemotherapy drugs reduces damage to healthy tissues and enhances therapeutic efficacy [63–65].

Chronic Disease Management: Biodegradable implants can release medications for conditions like diabetes or cardiovascular diseases over weeks or months, improving patient compliance.

Vaccination Platforms: Biodegradable materials are used to develop vaccine delivery systems that provide controlled release and improve immune responses [66].

By combining precision and biocompatibility, biodegradable drug delivery systems are transforming the way therapies are administered, ensuring better outcomes and reducing the burden on healthcare systems.

We explored the use of 3D printing to create biodegradable orthopedic implants capable of delivering antibiotics directly to surgical sites. The authors have utilized polylactic acid (PLA) as the base material, incorporating antibiotics including gentamicin into the PLA matrix. These antibiotic-infused PLA filaments were extruded and then 3D printed into various orthopedic device shapes, including screws, pins, and bone plates. To assess the performance of these constructs, the researchers conducted drug release studies using spectrophotometry over a one-week period, demonstrating sustained antibiotic elution. Mechanical properties were evaluated using tensile and compression testing, confirming that the printing process did not compromise the structural integrity of the implants. Biocompatibility and antibacterial efficacy were assessed through in vitro assays, which showed effective bacterial inhibition without adverse effects on mammalian cells. The study results show that 3D printing antibiotic-eluting, biodegradable orthopedic implants is a viable approach, offering the potential for personalized medicine solutions in orthopedic surgery. These custom-designed devices could provide tailored drug delivery and fit, enhancing patient-specific treatment options [36].

6 Applications in Tissue Engineering

Biodegradable materials, when combined with 3D printing technologies, have revolutionized tissue engineering by enabling the creation of scaffolds that support cellular growth and tissue regeneration. These scaffolds mimic the natural extracellular matrix, providing a conducive environment for cells to adhere, proliferate, and differentiate. This section explores the role of 3D printed biodegradable scaffolds in tissue engineering and their transformative impact on regenerative medicine.

6.1 Design and Fabrication of Scaffolds

The integration of 3D printing with biodegradable materials allows for the precise design and fabrication of scaffolds tailored to specific applications. Key features include:

Customizable Architectures: Scaffolds can be designed with intricate internal structures to mimic the mechanical and biological properties of natural tissues [67].

Material Versatility: A wide range of biodegradable polymers can be utilized, each tailored to specific degradation rates and mechanical requirements [68].

Integration with Bioactive Molecules: Scaffolds can be impregnated with growth factors [69], peptides [70], or other bioactive agents [36] to enhance cell adhesion, proliferation, and differentiation.

6.2 Application in Regenerative Medicine

Biodegradable scaffolds have found applications in various areas of regenerative medicine, including:

Bone and Cartilage Repair: Scaffolds provide structural support and promote the regeneration of bone and cartilage tissues, offering solutions for fractures, osteoarthritis, and other skeletal conditions [71]. In a research study conducted by Shen et al., an extrusion-based bioprinting technique was used to incorporate bone marrow-derived mesenchymal stem cells (MSCs) and rat aortic endothelial cells into a biodegradable PLA polymer matrix and 3D printed scaffolds to evaluate rat calvarial defect healing. Their study has shown that multiple cell lines can be successfully incorporated into the biodegradable polymer matrix and yielded enhanced growth compared to single cell polymer groups [72].

Cardiovascular Tissue Engineering: Scaffolds are used to create vascular grafts and patches for heart tissue repair, addressing conditions such as congenital defects or myocardial infarction [73].

Skin and Wound Healing: 3D printed biodegradable scaffolds are utilized in wound dressings and skin grafts, promoting faster and more effective healing [74].

As research advances, the development of multifunctional scaffolds with enhanced properties is becoming a focus. These include scaffolds with dynamic mechanical properties, on-demand drug release, and the ability to monitor healing processes. The convergence of biodegradable materials and 3D printing in tissue engineering continues to pave the way for novel treatments and improved patient outcomes.

7 Challenges and Future Directions

While the integration of biodegradable materials and 3D printing has revolutionized the biomedical sector, several challenges remain. Addressing these challenges is critical to fully unlocking the potential of this technology and ensuring its widespread adoption.

7.1 Challenges in Material Development

Below are several challenges regarding material development:

Limited Material Availability: The development of biodegradable materials that meet specific mechanical, biological, and degradation requirements remains a challenge [75].

Controlling Degradation Rates: Achieving precise control over the degradation kinetics of materials is essential to align with varying healing processes and medical applications [76].

Cost and Scalability: Producing biodegradable materials at scale while maintaining affordability and quality is an ongoing obstacle [77].

Regulatory Approvals and Ethical Concerns: The stringent regulatory landscape for medical devices and implants poses challenges for the adoption of novel biodegradable materials and 3D printed solutions [78]. Additionally, ensuring ethical practices in the sourcing and testing of biodegradable materials, particularly those derived from biological origins, is critical [79].

7.2 Future Research and Technological Advancement

With wide acceptability and increasing affordability the future advancements might look like:

Multifunctional Materials: Developing materials with integrated properties, such as antimicrobial effects or real-time monitoring capabilities, will expand their applications.

AI and Machine Learning: Utilizing AI to optimize scaffold designs and predict material behavior under physiological conditions can accelerate innovation [80].

Sustainability: Advancing eco-friendly manufacturing processes for biodegradable materials aligns with global sustainability goals.

By addressing these challenges and leveraging ongoing research, the integration of biodegradable materials and 3D printing will continue to transform the biomedical field, driving innovation and improving patient outcomes.

8 Conclusions

The convergence of biodegradable materials and 3D printing has set a new benchmark in the biomedical sector, enabling personalized, efficient, and sustainable healthcare solutions. These materials, with their unique properties and advantages, have found applications ranging from orthopedics and drug delivery systems to tissue engineering and regenerative medicine. The precision and adaptability of 3D printing further enhance the potential of these materials, offering tailored solutions that address individual patient needs.

Despite the transformative impact, challenges in material development, regulatory frameworks, and ethical considerations must be addressed to realize the full potential of this technology. Ongoing research and advancements in multifunctional materials, AI-driven design, and sustainable manufacturing hold promise for overcoming these hurdles.

As the biomedical field continues to evolve, the integration of biodegradable materials and 3D printing stands as a testament to the power of innovation in improving patient outcomes, reducing healthcare costs, and shaping the future of medicine.

As a 3D printing engineer and researcher working at the intersection of biomedical engineering and additive manufacturing, my work focuses on designing and fabricating biodegradable scaffolds that facilitate tissue regeneration, with a particular emphasis on bone and osteochondral interfaces. The crux of my research lies in understanding the interplay between material composition, architecture, and biological performance. Over the past several years, my team has developed custom 3D printing platforms capable of producing multi-material and gradient scaffolds that replicate the mechanical and biological gradients found in natural tissues. By employing materials such as poly(lactic-co-glycolic acid) (PLGA), polycaprolactone (PCL), and hydroxyapatite (HA), we have optimized scaffold structures to mimic trabecular bone architecture, ensuring both mechanical integrity and osteoconductivity.

Our research group's recent advances have focused on integrating bioactive molecules, stem cells, and responsive delivery systems within 3D printed constructs to enhance regeneration outcomes. For example, we have developed layered composite scaffolds using extrusion-based bioprinting techniques that can simultaneously support chondrogenesis and osteogenesis by creating zonal microenvironments. Moreover, leveraging AI-driven design algorithms, we have implemented real-time feedback systems that allow for adaptive printing based on porosity, density, and flow behavior of bioinks. This level of control has enabled us to tailor the degradation kinetics and mechanical strength of implants to specific patient anatomies, especially in pediatric and craniofacial applications. Currently, we are collaborating with clinical partners to translate these technologies into in vivo animal models and eventually human trials. Our aim is to bridge the gap between bench-top fabrication and functional, long-lasting, and biologically integrated implants in clinical practice.

References

1. Meng, M., Wang, J., Huang, H., Liu, X., Zhang, J., Li, Z.: 3D printing metal implants in orthopedic surgery: methods, applications and future prospects. J. Orthop. Translat. **42**, 94 (2023). https://doi.org/10.1016/J.JOT.2023.08.004

2. Alonzo, M., Alvarez Primo, F., Anil Kumar, S., Mudloff, J.A., Dominguez, E., Fregoso, G., Ortiz, N., Weiss, W.M., Joddar, B.: Bone tissue engineering techniques, advances and scaffolds for treatment of bone defects. Curr. Opin. Biomed. Eng. **17**, 100248 (2020). https://doi.org/10.1016/J.COBME.2020.100248
3. Bácskay, I., Ujhelyi, Z., Fehér, P., Arany, P.: The evolution of the 3D printed drug delivery systems: a review. Pharmaceutics. **14**, 1312 (2022). https://doi.org/10.3390/PHARMACEUTICS14071312
4. Zhao, X., Li, N., Zhang, Z., Hong, J., Zhang, X., Hao, Y., Wang, J., Xie, Q., Zhang, Y., Li, H., Liu, M., Zhang, P., Ren, X., Wang, X.: Beyond hype: unveiling the real challenges in clinical translation of 3D printed bone scaffolds and the fresh prospects of bioprinted organoids. J. Nanobiotechnol. **22**(1), 1–26 (2024). https://doi.org/10.1186/S12951-024-02759-Z
5. Barcena, A.J.R., Ravi, P., Kundu, S., Tappa, K.: Emerging biomedical and clinical applications of 3D printed poly(Lactic Acid)-based devices and delivery systems. Bioengineering **11**, 705 (2024). https://doi.org/10.3390/BIOENGINEERING11070705
6. Lewandrowski, K.U., Vira, S., Elfar, J.C., Lorio, M.P.: Advancements in custom 3D printed titanium interbody spinal fusion cages and their relevance in personalized spine care. J. Pers. Med. **14**, 809 (2024). https://doi.org/10.3390/JPM14080809
7. Copeland, C., Reyes, C.C., Peck, J.L., Srivastava, R., Zuniga, J.M.: Functional performance and patient satisfaction comparison between a 3D printed and a standard transradial prosthesis: a case report. Biomed. Eng. Online **21**, 7 (2022). https://doi.org/10.1186/S12938-022-00977-W
8. Rezaie, F., Farshbaf, M., Dahri, M., Masjedi, M., Maleki, R., Amini, F., Wirth, J., Moharamzadeh, K., Weber, F.E., Tayebi, L.: 3D printing of dental prostheses: current and emerging applications. J. of Compos. Sci. **7**, 80 (2023). https://doi.org/10.3390/JCS7020080
9. Liang, W., Zhou, C., Zhang, H., Bai, J., Jiang, B., Jiang, C., Ming, W., Zhang, H., Long, H., Huang, X., Zhao, J.: Recent advances in 3D printing of biodegradable metals for orthopaedic applications. J. Biol. Eng. **17**, 56 (2023). https://doi.org/10.1186/S13036-023-00371-7
10. Domsta, V., Hänsch, C., Lenz, S., Gao, Z., Matin-Mann, F., Scheper, V., Lenarz, T., Seidlitz, A.: The influence of shape parameters on unidirectional drug release from 3D printed implants and prediction of release from implants with individualized shapes. Pharmaceutics **15**, 1276 (2023). https://doi.org/10.3390/PHARMACEUTICS15041276
11. Li, L., Li, Y., Yang, L., Yu, F., Zhang, K., Jin, J., Shi, J., Zhu, L., Liang, H., Wang, X., Jiang, Q.: Polydopamine coating promotes early osteogenesis in 3D printing porous Ti6Al4V scaffolds. Ann Transl Med. **7**, 240 (2019). https://doi.org/10.21037/ATM.2019.04.79
12. Damle, M.N., Chaudhari, L., Tardalkar, K., Bhamare, N., Jagdale, S., Gaikwad, V., Chhabra, D., Kumar, B., Manuja, A., Joshi, M.G.: A biologically functional bioink based on extracellular matrix derived collagen for 3D printing of skin. Int. J. Biol. Macromol. **258**, 128851 (2024). https://doi.org/10.1016/J.IJBIOMAC.2023.128851
13. Dinh, L., Machamasi, R., Kim, C.J., Lee, J.J., Choi, Y., Kim, H., Mahon, L., Grabenbauer, A., Yan, B., Hwang, S.J.: 3D printed chitosan-pectin-sodium alginate scaffolds for post-surgical peritoneal wound dressing and sustained delivery of oxaliplatin. J. Pharm. Investig. **54**, 857–871 (2024). https://doi.org/10.1007/S40005-024-00704-8/METRICS
14. Demirel, G., Cakıl, Y.D., Koltuk, G., Aktas, R.G., Calıskan, M.: The use of hyaluronic acid in a 3D biomimetic scaffold supports spheroid formation and the culture of cancer stem cells. Sci. Rep. **14**(1), 1–12 (2024). https://doi.org/10.1038/s41598-024-69047-6
15. Gregor, A., Filová, E., Novák, M., Kronek, J., Chlup, H., Buzgo, M., Blahnová, V., Lukášová, V., Bartoš, M., Nečas, A., Hošek, J.: Designing of PLA scaffolds for bone tissue replacement fabricated by ordinary commercial 3D printer. J. Biol. Eng. **11**, 1–21 (2017). https://doi.org/10.1186/S13036-017-0074-3/TABLES/4
16. Bassand, C., Siepmann, F., Benabed, L., Verin, J., Freitag, J., Charlon, S., Soulestin, J., Siepmann, J.: 3D printed PLGA implants: how the filling density affects drug release. J. Control. Release **363**, 1–11 (2023). https://doi.org/10.1016/J.JCONREL.2023.09.020
17. Wattanaanek, N., Suttapreyasri, S., Samruajbenjakun, B.: 3D printing of calcium phosphate/calcium sulfate with alginate/cellulose-based scaffolds for bone regeneration: multilayer fabrication and characterization. J. Funct. Biomater. **13** (2022). https://doi.org/10.3390/JFB13020047

18. Rajabi, M., Cabral, J.D., Saunderson, S., Ali, M.A.: 3D printing of chitooligosaccharide-polyethylene glycol diacrylate hydrogel inks for bone tissue regeneration. J. Biomed. Mater. Res. A **111**, 1468–1481 (2023). https://doi.org/10.1002/JBM.A.37548
19. Beheshtizadeh, N., Farzin, A., Rezvantalab, S., Pazhouhnia, Z., Lotfibakhshaiesh, N., Ai, J., Noori, A., Azami, M.: 3D printing of complicated GelMA-coated Alginate/Tri-calcium silicate scaffold for accelerated bone regeneration. Int. J. Biol. Macromol. **229**, 636–653 (2023). https://doi.org/10.1016/J.IJBIOMAC.2022.12.267
20. Sun, T., Feng, Z., He, W., Li, C., Han, S., Li, Z., Guo, R.: Novel 3D printing bilayer GelMA-based hydrogel containing BP, β-TCP and exosomes for cartilage-bone integrated repair. Biofabrication **16** (2023). https://doi.org/10.1088/1758-5090/AD04FE
21. Guerra, A.J., Cano, P., Rabionet, M., Puig, T., Ciurana, J.: 3D printed PCL/PLA composite stents: towards a new solution to cardiovascular problems. Materials (Basel) **11** (2018). https://doi.org/10.3390/MA11091679
22. Babilotte, J., Martin, B., Guduric, V., Bareille, R., Agniel, R., Roques, S., Héroguez, V., Dussauze, M., Gaudon, M., Le Nihouannen, D., Catros, S.: Development and characterization of a PLGA-HA composite material to fabricate 3D printed scaffolds for bone tissue engineering. Mater. Sci. Eng. C Mater. Biol. Appl. **118** (2021). https://doi.org/10.1016/J.MSEC.2020.111334
23. Suo, H., Zhang, J., Xu, M., Wang, L.: Low-temperature 3D printing of collagen and chitosan composite for tissue engineering. Mater. Sci. Eng. C Mater. Biol. Appl. **123** (2021). https://doi.org/10.1016/J.MSEC.2021.111963
24. Cresswell-Boyes, A.J., Davis, G.R., Krishnamoorthy, M., Mills, D., Barber, A.H.: Composite 3D printing of biomimetic human teeth. Sci Rep. **12** (2022). https://doi.org/10.1038/S41598-022-11658-Y
25. Tappa, K., Jammalamadaka, U., Ballard, D.H., Bruno, T., Israel, M.R., Vemula, H., Meacham, J.M., Mills, D.K., Woodard, P.K., Weisman, J.A.: Medication eluting devices for the field of OBGYN (MEDOBGYN): 3D printed biodegradable hormone eluting constructs, a proof of concept study. PLoS ONE **12**, e0182929 (2017). https://doi.org/10.1371/journal.pone.0182929
26. Joseph Rey, J.R.H., Chen, Q., Maalihan, R.D., Ren, J., da Silva, Í.G.M., Dugos, N.P., Caldona, E.B., Advincula, R.C.: 3D printing of biomedically relevant polymer materials and biocompatibility. MRS Commun. **11**(2), 197–212 (2021). https://doi.org/10.1557/S43579-021-00038-8
27. Makadia, H.K., Siegel, S.J.: Poly lactic-co-glycolic acid (PLGA) as biodegradable controlled drug delivery carrier. Polymers (Basel) **3**, 1377 (2011). https://doi.org/10.3390/POLYM3031377
28. Mirkhalaf, M., Men, Y., Wang, R., No, Y., Zreiqat, H.: Personalized 3D printed bone scaffolds: a review. Acta Biomater. **156**, 110–124 (2023). https://doi.org/10.1016/J.ACTBIO.2022.04.014
29. Fidanza, A., Perinetti, T., Logroscino, G., Saracco, M.: 3D printing applications in orthopaedic surgery: clinical experience and opportunities. Appl. Sci. **12**, 3245 (2022). https://doi.org/10.3390/APP12073245
30. Rocha, C.V., Gonçalves, V., da Silva, M.C., Bañobre-López, M., Gallo, J.: PLGA-based composites for various biomedical applications. Int. J. Mol. Sci. **23**, 2034 (2022). https://doi.org/10.3390/IJMS23042034
31. Ravi, P., Burch, M.B., Farahani, S., Chepelev, L.L., Yang, D., Ali, A., Joyce, J.R., Lawera, N., Stringer, J., Morris, J.M., Ballard, D.H., Wang, K.C., Mahoney, M.C., Kondor, S., Rybicki, F.J., Rabinowitz, Y.A., Shapiro, S.B., McCormick, B., Costea, A.I., Byrd, S., Panza, A., Danesi, T.H., Giglia, J.S., Chadalavada, S., Krishnan, D.G., Cervenka, B.P., Phero, J.A., McLaurin, W.S., Sidana, A., Utz, C.J., Grawe, B.: Utility and costs during the initial year of 3D printing in an Academic Hospital. J. Am. Coll. Radiol. **20**, 193–204 (2023). https://doi.org/10.1016/J.JACR.2022.07.001
32. Nudelman, H., Lőrincz, A., Lamberti, A.G., Varga, M., Kassai, T., Józsa, G.: Management of pediatric ankle fractures: comparison of biodegradable PLGA implants with traditional metal screws. Front. Pediatr. **12**, 1410750 (2024). https://doi.org/10.3389/FPED.2024.1410750/FULL

33. Åkerlund, E., Diez-Escudero, A., Grzeszczak, A., Persson, C.: The effect of PCL addition on 3D printable PLA/HA composite filaments for the treatment of bone defects. Polymers (Basel) **14** (2022). https://doi.org/10.3390/polym14163305
34. Bernardo, M.P., da Silva, B.C.R., Hamouda, A.E.I., de Toledo, M.A.S., Schalla, C., Rütten, S., Goetzke, R., Mattoso, L.H.C., Zenke, M., Sechi, A.: PLA/Hydroxyapatite scaffolds exhibit in vitro immunological inertness and promote robust osteogenic differentiation of human mesenchymal stem cells without osteogenic stimuli. Sci. Rep. **12** (2022). https://doi.org/10.1038/s41598-022-05207-w
35. Wu, D., Spanou, A., Diez-Escudero, A., Persson, C.: 3D printed PLA/HA composite structures as synthetic trabecular bone: a feasibility study using fused deposition modeling. J. Mech. Behav. Biomed. Mater. **103** (2020). https://doi.org/10.1016/j.jmbbm.2019.103608
36. Wu, L., Shrestha, P., Iapichino, M., Cai, Y., Kim, B., Stoeber, B.: Characterization method for calculating diffusion coefficient of drug from polylactic acid (PLA) microneedles into the skin. J. Drug Deliv. Sci. Technol. **61** (2021). https://doi.org/10.1016/j.jddst.2020.102192
37. Tappa, K., Jammalamadaka, U., Weisman, J.A., Ballard, D.H., Wolford, D.D., Pascual-Garrido, C., Wolford, L.M., Woodard, P.K., Mills, D.K.: 3D printing custom bioactive and absorbable surgical screws, pins, and bone plates for localized drug delivery. J. Funct. Biomater. **10** (2019). https://doi.org/10.3390/jfb10020017
38. Băilă, D.I., Tonoiu, S.: Properties of photo-curable polyurethane resins used in SLA manufacturing. IOP Conf. Ser. Mater. Sci. Eng. **1268**, 012006 (2022). https://doi.org/10.1088/1757-899x/1268/1/012006
39. Revilla-León, M., Meyers, M.J., Zandinejad, A., Özcan, M.: A review on chemical composition, mechanical properties, and manufacturing work flow of additively manufactured current polymers for interim dental restorations. (2019)
40. Creylman, V., Muraru, L., Pallari, J., Vertommen, H., Peeraer, L.: Gait assessment during the initial fitting of customized selective laser sintering ankle foot orthoses in subjects with drop foot. Prosthet. Orthot. Int. **37**, 132–138 (2013). https://doi.org/10.1177/0309364612451269
41. Li, N., Guo, R., Zhang, Z.J.: Bioink formulations for bone tissue regeneration. (2021)
42. Gillispie, G.J., Copus, J., Uzun-Per, M., Yoo, J.J., Atala, A., Niazi, M.K.K., Lee, S.J.: The correlation between rheological properties and extrusion-based printability in bioink artifact quantification. Mater. Des. 233 (2023). https://doi.org/10.1016/j.matdes.2023.112237
43. Hu, Y., Luo, Z., Bao, Y.: Trends in photopolymerization 3D printing for advanced drug delivery applications. Biomacromolecules **26**, 85 (2024). https://doi.org/10.1021/ACS.BIOMAC.4C01004
44. Bergeron, L., Bonapace-Potvin, M., Bergeron, F.: In-house 3D model printing for acute cranio-maxillo-facial trauma surgery: process, time, and costs. Plast. Reconstr. Surg. Glob. Open **9**, E3804 (2021). https://doi.org/10.1097/GOX.0000000000003804
45. Veerubhotla, K., Lee, C.H.: Design of biodegradable 3D printed cardiovascular stent. Bioprinting **26** (2022). https://doi.org/10.1016/j.bprint.2022.e00204
46. Raja, N., Han, S.H., Cho, M., Choi, Y.J., Jin, Y.Z., Park, H., Lee, J.H., Yun, H.S.: Effect of porosity and phase composition in 3D printed calcium phosphate scaffolds on bone tissue regeneration in vivo. Mater. Des. **219** (2022). https://doi.org/10.1016/j.matdes.2022.110819
47. Mills, D., Tappa, K., Jammalamadaka, U., Weisman, J., Woerner, J.: The use of 3D printing in the fabrication of nasal stents. Inventions **3**, 1 (2017). https://doi.org/10.3390/inventions3010001
48. Wong, R.M.Y., Wong, P.Y., Liu, C., Chung, Y.L., Wong, K.C., Tso, C.Y., Chow, S.K.H., Cheung, W.H., Yung, P.S.H., Chui, C.S., Law, S.W.: 3D printing in orthopaedic surgery: A scoping review of randomized controlled trials. Bone Joint Res. **10** (2021). https://doi.org/10.1302/2046-3758.1012.BJR-2021-0288.R2
49. Zhang, B., Sun, H., Wu, L., Ma, L., Xing, F., Kong, Q., Fan, Y., Zhou, C., Zhang, X.: 3D printing of calcium phosphate bioceramic with tailored biodegradation rate for skull bone tissue reconstruction. Biodes. Manuf. **2** (2019). https://doi.org/10.1007/s42242-019-00046-7
50. Ali, F., Kalva, S.N., Koç, M.: Additive manufacturing of polymer/Mg-based composites for porous tissue scaffolds. (2022)

51. Slavin, B.V., Ehlen, Q.T., Costello, J.P., Nayak, V.V., Bonfante, E.A., Benalcázar Jalkh, E.B., Runyan, C.M., Witek, L., Coelho, P.G.: 3D printing applications for craniomaxillofacial reconstruction: a sweeping review. (2023)
52. Liu, Y., Wu, H., Bao, S., Huang, H., Tang, Z., Dong, H., Liu, J., Chen, S., Wang, N., Wu, Z., Zhang, Z., Shi, L., Li, X., Guo, Z.: Clinical application of 3D printed biodegradable lumbar interbody cage (polycaprolactone/β-tricalcium phosphate) for posterior lumbar interbody fusion. J. Biomed. Mater. Res. B Appl. Biomater. **111** (2023). https://doi.org/10.1002/jbm.b.35244
53. Yang, L., Fan, L., Lin, X., Yu, Y., Zhao, Y.: Pearl powder hybrid bioactive scaffolds from microfluidic 3D printing for bone regeneration. Adv. Sci. **10** (2023). https://doi.org/10.1002/advs.202304190
54. Pan, T., Song, W., Xin, H., Yu, H., Wang, H., Ma, D., Cao, X., Wang, Y.: MicroRNA-activated hydrogel scaffold generated by 3D printing accelerates bone regeneration. Bioact. Mater. **10**, 1–14 (2022). https://doi.org/10.1016/j.bioactmat.2021.08.034
55. Jammalamadaka, U., Tappa, K., Mills, D.K.: Calcium phosphate/clay nanotube bone cement with enhanced mechanical properties and sustained drug release. In: Current Topics in the Utilization of Clay in Industrial and Medical Applications. InTech (2018)
56. Jammalamadaka, U., Tappa, K., Mills, D.: Osteoinductive calcium phosphate clay nanoparticle bone cements (CPCs) with enhanced mechanical properties. Conf. Proc. IEEE Eng. Med. Biol. Soc. **2014**, 3917–3920 (2014). https://doi.org/10.1109/EMBC.2014.6944480
57. Tappa, K., Jammalamadaka, U., Mills, D.K.: Formulation and evaluation of nano enhanced anti-bacterial calcium phosphate bone cements. In: Li, B., Webster, T. (eds.) Orthopedic biomaterials: advances and applications, pp. 85–108. Springer International Publishing, Cham (2017)
58. Weisman, J., Jammalamadaka, U., Tappa, K., Mills, D.: Doped halloysite nanotubes for use in the 3D printing of medical devices. Bioengineering **4**, 96 (2017). https://doi.org/10.3390/bioengineering4040096
59. Mills, D.K., Jammalamadaka, U., Tappa, K., Weisman, J.: Studies on the cytocompatibility, mechanical and antimicrobial properties of 3D printed poly(methyl methacrylate) beads. Bioact Mater. **3**, 157–166 (2018). https://doi.org/10.1016/J.BIOACTMAT.2018.01.006
60. Youssef, S.H., Ganesan, R., Amirmostofian, M., Kim, S., Polara, R., Afinjuomo, F., Song, Y., Chereda, B., Singhal, N., Robinson, N., Garg, S.: Printing a cure: a tailored solution for localized drug delivery in liver cancer treatment. Int. J. Pharm. 651. https://doi.org/10.1016/j.ijpharm.2024.123790
61. Loh, J.M., Lim, Y.J.L., Tay, J.T., Cheng, H.M., Tey, H.L., Liang, K.: Design and fabrication of customizable microneedles enabled by 3D printing for biomedical applications. (2024)
62. Zhang, P., Li, J., Ashour, E.A., Chung, S., Wang, H., Vemula, S.K., Repka, M.A.: Development of multiple structured extended release tablets via hot melt extrusion and dual-nozzle fused deposition modeling 3D printing. Int. J. Pharm. **653** (2024). https://doi.org/10.1016/j.ijpharm.2024.123905
63. Wei, H., Luo, Y., Ma, R., Li, Y.: Three-dimensional printing multi-drug delivery core/shell fiber systems with designed release capability. Pharmaceutics **15** (2023). https://doi.org/10.3390/pharmaceutics15092336
64. Cho, H., Jammalamadaka, U., Tappa, K.: Nanogels for pharmaceutical and biomedical applications and their fabrication using 3D printing technologies. Materials **11** (2018). https://doi.org/10.3390/ma11020302
65. Weisman, J.A., Nicholson, J.C., Tappa, K., Jammalamadaka, U., Wilson, C.G., Mills, D.K.: Antibiotic and chemotherapeutic enhanced three-dimensional printer filaments and constructs for biomedical applications. Int. J. Nanomedicine **10**, 357–370 (2015). https://doi.org/10.2147/IJN.S74811
66. Weisman, J.A., Ballard, D.H., Jammalamadaka, U., Tappa, K., Sumerel, J., D'Agostino, H.B., Mills, D.K., Woodard, P.K.: 3D printed antibiotic and chemotherapeutic eluting catheters for potential use in interventional radiology: in vitro proof of concept study. Acad. Radiol. **26** (2019). https://doi.org/10.1016/j.acra.2018.03.022

67. Varghese, R., Salvi, S., Sood, P., Karsiya, J., Kumar, D.: 3D printed medicine for the management of chronic diseases: the road less travelled.
68. Zhou, J., Li, Q., Tian, Z., Yao, Q., Zhang, M.: Recent advances in 3D bioprinted cartilage-mimicking constructs for applications in tissue engineering. (2023)
69. Tetsuka, H., Shin, S.R.: Materials and technical innovations in 3D printing in biomedical applications. (2020)
70. Bittner, S.M., Guo, J.L., Mikos, A.G.: Spatiotemporal control of growth factors in three-dimensional printed scaffolds. (2018)
71. Hammerstone, D.E., Babuska, T.F., Lazarte, S., Krick, B.A., Chow, L.W.: Characterizing properties of scaffolds 3D printed with peptide-polymer conjugates. Biomater Adv **152** (2023). https://doi.org/10.1016/j.bioadv.2023.213498
72. Xu, N., Lu, D., Qiang, L., Liu, Y., Yin, D., Wang, Z., Luo, Y., Yang, C., Ma, Z., Ma, H., Wang, J.: 3D printed composite bioceramic scaffolds for bone and cartilage integrated regeneration. ACS Omega **8** (2023). https://doi.org/10.1021/acsomega.3c03284
73. Shen, M., Wang, L., Gao, Y., Feng, L., Xu, C., Li, S., Wang, X., Wu, Y., Guo, Y., Pei, G.: 3D bioprinting of in situ vascularized tissue engineered bone for repairing large segmental bone defects. Mater. Today Bio. **16** (2022). https://doi.org/10.1016/j.mtbio.2022.100382
74. Wang, Z., Wang, L., Li, T., Liu, S., Guo, B., Huang, W., Wu, Y.: 3D bioprinting in cardiac tissue engineering. (2021)
75. Uchida, D.T., Bruschi, M.L.: 3D printing as a technological strategy for the personalized treatment of wound healing. (2023)
76. Iftekar, S.F., Aabid, A., Amir, A., Baig, M.: Advancements and limitations in 3D printing. Mater Technol A Crit Rev
77. Locks, A., Bowles, B.J., Brown, S., Hailes, H.C., Hilton, S.T.: 3D printing with tuneable degradation: thiol-ene and thiol-yne containing formulations for biomedical applications. Int. J. Pharm. **674**, 125432 (2025). https://doi.org/10.1016/J.IJPHARM.2025.125432
78. Kim, M., Yoo, G., Kim, B., Song, Y., Lee, B.J.: Scalability enhancement in projection-based 3D printing through optical expansion. Addit. Manuf. **95**, 104511 (2024). https://doi.org/10.1016/J.ADDMA.2024.104511
79. Mladenovska, T., Choong, P.F., Wallace, G.G., O'connell, C.D.: The regulatory challenge of 3D bioprinting. (2023)
80. Neely, E.L.: The risks of revolution: ethical dilemmas in 3D printing from a US Perspective. Sci. Eng. Ethics **22** (2016). https://doi.org/10.1007/s11948-015-9707-4

Karthik Tappa , PhD is the 3D printing engineer in the MD Anderson Cancer Center's Anatomic 3D Printing and Visualization Program. As a biomedical engineer specialized in multiple additive manufacturing technologies, he has made significant contributions to the field of medical 3D printing. His research has been published in numerous leading biomedical engineering journals, showcasing his innovative approaches to personalized healthcare solutions. His work has not only expanded the theoretical understanding of 3D printing applications but has also translated into tangible advancements in patient care. Armed with a wealth of knowledge and expertise, Dr. Tappa has actively engaged with both academic and industry professionals. He has collaborated on interdisciplinary projects, fostering partnerships between engineering and medical communities. As a visionary in the field of point-of-care 3D printing, Dr. Tappa continues to explore new frontiers and envision a future where personalized medicine is enhanced through the seamless integration of engineering innovation.

Chapter 9
Wet-Mold Concrete Casting with Biodegradable Materials

Brendan Kiser, Ryan Kennedy, and Lucas Palicki

Abstract Recent trends in the construction industry are asking for greater flexibility and control of free-form concrete geometries, either for an environmentally conscious material reduction or for aesthetic purposes. Complex concrete components must be economically fabricated to a high degree of precision, where especially the cost of the formwork can become prohibitively expensive. We therefore propose the formwork for individually shaped concrete components to be made of recyclable wood materials by means of 3D printing and wet-mold casting. The developed method uses fully biodegradable formwork and has been tested for the design of unique building facade units. Shrinkage has been identified as a common problem in the large-scale 3D printing of fully biodegradable materials such as wood or cellulose. To sidestep this limitation of organic material 3D printing, we propose a wet-mold casting whereby the concrete cures prior to the drying of the mold material. A secondary rectangular exterior mold withstands the hydrostatic pressure. The design of the desired geometry is sliced into contours. Upon deposition of each layer of wood material into the exterior mold by a robotic arm, further wood material is used to fill the outside of the mold, while concrete is fed into the space inside. The concrete is left to cure, with the wet-mold wood material removed while it is still malleable. The wood material can be reused or left to biodegrade. The possibilities of this process were explored in the design of a building with individual structural units. The developed method, which utilizes reclaimable mold material, may offer a pathway toward mass-customization of concrete components at an architectural scale, potentially allowing each unit in a structure to be unique with minimal added cost.

Keywords 3D printing · Wet-mold casting · Concrete · Sustainability · Formwork

B. Kiser (✉)
College of Art and Design, Southern Illinois University, Carbondale, IL, USA
e-mail: bkiser630@gmail.com

R. Kennedy
Champlin/EOP Architects, Cincinnati, OH, USA

L. Palicki
Palicki Golf Design, Cincinnati, OH, USA

© The Author(s), under exclusive license to Springer Nature Singapore Pte Ltd. 2026
C. Klemmt and G. Beaucage (eds.), *Biodegradable 3D-Printing*, The Materials Research Society Series, https://doi.org/10.1007/978-981-95-4025-9_9

1 Introduction

The construction industry faces increasing demand for more sustainable and efficient methods while accommodating increasingly complex and customized architectural designs. Traditional construction techniques, constrained by manual processes and material limitations, may struggle to meet these requirements. Some of these methods, if applied to complex or free-form geometries, result in material waste, high costs, and limited flexibility in the overall design. To address these challenges, modern fabrication methods such as 3D printing are being integrated into building design and construction, offering a transformative approach to how structures are created.

3D printing is often highlighted for its potential advantages such as greater design flexibility, reduced material waste, and faster production of detailed components at lower costs, particularly for custom and intricate designs. In some cases, 3D printing is even employed to construct entire buildings from the ground up using large-scale 3D printers. This approach enables architects and builders to push the boundaries of conventional design, utilizing advanced tools like parametric modeling and topology optimization. These methods ensure structural integrity while maximizing material efficiency, resulting in innovative and resource-conscious designs that were previously impractical or impossible to achieve.

One material relevant to this discussion is concrete. Though widely used in construction for its strength, durability, and versatility, it is sometimes considered difficult for nontraditional designs due to factors such as its mass, reinforcement requirements, and the costs associated with producing custom molds. Combining 3D printing with concrete production provides an opportunity to address previous limitations, enabling more customized and efficient building components while minimizing material waste. This approach enables the production of complex forms that were previously considered difficult and costly to achieve.

While there are many advantages to 3D printed concrete, one of its limitations is often considered to be the difficulty in reinforcing the printed structure. Printed materials can lack the tensile strength needed for application. However, this issue may be bypassed by using 3D printing to create molds for casting concrete, providing a product that has compressive strength which is still traditionally reinforced with steel rebar. By printing custom molds, designers can achieve precise, detailed shapes for concrete structures without compromising strength or durability. This approach combines the flexibility of 3D printing with the strength of traditional materials, overcoming the structural challenges posed by directly 3D printed structures.

To create entirely unique building components using this method, equally unique 3D printed molds are required as well: Each new concrete cast requires a new 3D printed mold. Although it is not currently possible to mass-recycle traditional thermoplastic or silicone molds in this manner, it is feasible with materials that can remain uncured during the casting process. We therefore propose using 3D printed molds for concrete casting, where the mold, made from a biodegradable wood material, remains in an uncured state when the concrete is poured. The printing path is generated based on the negative space of the desired form, with each layer of printed material stacking on the previous one. The mold is printed into an outer container that withstands the hydrostatic pressure of the concrete. As the 3D mold is printing, an outer layer of material is infilled between the 3D printed mold and the outer container, while the concrete is poured into the mold with reinforcement being placed as required. The concrete cures while the mold remains uncured, allowing for a reuse of the wood mold and infill material as soon as the concrete is self-supporting. Following removal of the 3D printed mold, the concrete unit depicts the texture and patterns of the biodegradable print (Fig. 1).

The ability to recycle and reuse the printed material greatly reduces cost and time required to produce the desired component at large scales. Moreover, the process is environmentally friendly, as the ingredients for the printed material are all fully biodegradable, avoiding any kind of waste. Thus, we propose a method that could potentially reduce environmental impact, construction costs, and material waste while enabling greater architectural flexibility.

Fig. 1 Final concrete unit following extraction from the biodegradable and recyclable mold. Note the embedded wood chips and the rough texture with visible layer lines

2 Related Work

2.1 Alternatives to Standard Concrete Construction

The comparison between standard concrete construction and alternative methods, such as 3D printed concrete, continues to attract interest in the construction industry. After an analysis was conducted on 3D printed concrete, Schutter and his team suggest that digitally fabricated concrete structures have the potential to outperform traditional concrete methods, especially when analyzing environmental sustainability over the structure's lifecycle [1]. This is particularly promising, as 3D printed concrete, when combined with supplementary binders such as fly ash [2, 3], silica fume [3, 4], and slag [2, 3], can enhance its sustainability potential. The supplementary materials, which are often by-products of industrial processes, can partially replace traditional cement and reduce the environmental impact associated with concrete building construction. Moreover, with the appropriate mixture of materials, 3D printed concrete can maintain the necessary mechanical properties, such as strength and durability, while offering a more sustainable solution as described by Kaszynska [5]. As 3D printing continues to evolve, it presents a compelling alternative to traditional construction methods by allowing for more sustainable and customizable designs.

In addition to 3D printing, other alternative methods for concrete construction have been noted for their ability to improve efficiency and sustainability. One such innovation is Smart Dynamic Casting, which involves a robotically controlled formwork system that moves vertically using a linear axis system [6]. Enabling more precise and efficient pouring of concrete around pre-installed reinforcement, this system increases construction accuracy while reducing waste. Another notable advancement is the use of robotically controlled arms to apply shotcrete, a modified form of concrete that is sprayed onto surfaces [7]. Together, these alternative construction methods provide significant enhancements to traditional concrete construction, offering greater efficiency, design flexibility, and environmental sustainability.

2.2 Geometry Optimized Concrete to Reduce CO2 Emissions

Topology-optimized concrete is described as a computational approach aimed at determining efficient material usage and structural performance for specific designs. Adapting this approach to building construction leads to more sustainable designs by reducing material waste and minimizing carbon footprint of structures. Researchers like Jipa [8, 9] have tested and demonstrated the effectiveness of this method, showcasing its potential for more sustainable architectural and engineering solutions. Innovations in geometry optimization, such as combining 3D printed concrete with topology optimization, further enhance its sustainability potential [10]. As displayed by Yang [10], this integration allows for the creation of intricate and efficient structures that maximize strength while minimizing material usage, ultimately contributing to significant reductions in CO2 emissions in the construction industry.

2.3 3D Printing in Architecture

3D printing has become an innovative tool in architecture, allowing for the printing of structures [11, 12], facades [13, 14], and formwork for casting of other materials [15, 16]. This technology enables architects to design and manufacture complex geometries with greater precision and efficiency than traditional methods. Additionally, the use of coding software allows for the exploration of computer-driven designs, enabling architects to push the boundaries of both form and structure as demonstrated by Dillenburger [17]. This integration of 3D printing with digital design tools offers a new level of creative freedom and potential for architecture.

While offering innovative solutions in manufacturing, 3D printing faces several challenges that can impact its efficiency and effectiveness as outlined by Ko [18]. One challenge is the dimension of the printing machine, which limits the size of the objects that can be printed. Additionally, the choice of robot type—articulated, cartesian, cylindrical, polar, selective compliance assembly robot arm, or delta—introduces inherent limitations in movement precision, affecting the final print quality. The nozzles used in the extrusion process play a key role in determining the print's width and shape, directly influencing its accuracy. The extrusion mechanism itself also presents challenges related to how material flows through the nozzle, affecting both the strength and form of the print. Additionally, material properties, such as slump rate, can impact the structural integrity of the print, with variations in consistency leading to errors in the final result. Furthermore, the manufacturing process has its own inherent limitations, including the difficulty of printing overhanging elements, the compression of layers which can affect the strength of the final structure, and the challenge of bridging 3D printed materials without a proper support structure. The stability of the print before it fully dries is another issue, as premature movement can lead to deformities or structural failure. These various factors contribute to the complexity of 3D printing, requiring careful consideration and precise control.

3D printed architecture offers significant potential for reducing the environmental impact of construction compared to traditional methods. Highlighted by Khan are several studies that explore the addition of more sustainable materials to 3D printed concrete, with many showing promising results for reducing the environmental impact [19]. While still being researched by Abdalla [20], Zhao [21], and others [19, 22], 3D printed structures are generally considered to be cheaper to produce than traditional ones, with further research necessary in order to confirm the cost-effectiveness in long term construction. Beyhan outlines the key benefits of 3D printing which include affordable customization, a single machine with limitless product range, and a more efficient approach to raw material usage, contributing to a reduction in waste when compared to traditional construction methods [23].

However, 3D printing of architecture still faces many challenges, including the high cost and limited availability of the technology, the absence of standardized building codes, and a general lack of industry demand and skill. Despite these obstacles, the potential for 3D printing to transform sustainable construction practices is significant and should be further explored.

2.4 3D Printing of Biodegradable Materials

The field of 3D printing with biodegradable materials is experiencing significant advancements through various studies focused on optimizing both material properties and printing techniques. A variety of biodegradable materials are being tested for 3D printing use such as wood composite powders [24, 25], lignin [26, 27], cellulose [26, 27], and spruce wood chips [28, 29], combined with xanthan gum [26], methylcellulose [26, 28, 30], polylactic acid (PLA) [31], and other binders. By adjusting these ratios, researchers can better understand how to achieve optimal strength for a range of applications. Additionally, researchers are investigating fiber layouts in printed wood to control shrinkage during drying [32], how print depth affects the 3D prints strength and surface quality [33], and examining the impact of print speed on mechanical properties of the various results [34]. Together, these studies offer valuable insights into optimizing the 3D printing process for biodegradable wood materials, with the goal of achieving more consistent and sustainable results.

Recent studies have highlighted the diverse applications of 3D printed biodegradable materials, with a particular focus on wood-based composites. Three primary uses for 3D printed wood have been identified by Wimmer [35]: The development of bio-based filaments [27, 31], the printing of wood-like 3D objects [31], and the "self-replication" of wood through the combination of 3D printing and microtomography [36]. In one such study, particleboard was broken down to a printable scale, then mixed with PLA to create an extrudable filament. This filament was analyzed and then used to print a functional prototype, a desk table, showcasing the potential of 3D printed wood materials for sustainable applications [31]. Another study used a mixture of lignin and cellulose to create a filament, allowing for the creation of designed structures. After heat treatment, the resulting material closely resembles the properties of 3D printed wood designs, enhancing its appeal as a viable alternative to traditional wood products [27]. These advancements reflect the increasing potential of biodegradable 3D printed materials to replace conventional methods, offering both environmental and functional benefits in many applications.

2.5 3D Printed Molds for Concrete Casting

The exploration of 3D printed biodegradable molds for concrete casting focuses on the potential for sustainable formwork solutions in construction. Several 3D printing processes are evaluated by Jipa, such as material extrusion of concrete, clay and polymers [15, 37, 38], material jetting of wax or foam [39, 40], and binder jetting of sand and binder [8, 9], testing their ability to create eco-friendly molds [41]. Other projects such as Earthworks investigates the viability of using dirt-based formworks as a temporary mold for concrete casting [42]. This approach reduces reliance on traditional mold materials and promotes the use of eco-friendly natural resources. Another notable example is the development of a 3D printed polyvinyl alcohol (PVA) shell by Leschok, which is fully biodegradable and dissolves with water after the concrete has cured [43]. Additionally, a 3D wood-based formwork has been designed by Khan to be printed, then ground down once the concrete is set [15]. While still under testing, Khan is proposing the ground material would be rehydrated and reused in future prints to create a closed-loop system, reducing waste and promoting more sustainable practices [15].

In contrast, non-biodegradable molds have been widely used in the 3D printing of concrete formworks due to their durability and reusability. A notable advancement in this area is the combination of binder jetting and fused deposition modeling (FDM), as explored by Jipa [9], which allows for the production of intricate and precise plastic molds. These molds can withstand the pressures of the concrete casting process and can be reused multiple times, ideal for large-scale projects. Another example demonstrates the use of 3D printed plastic formworks for concrete casting, removing the formwork after the concrete cures using a heat gun and pliers [37]. This process was later refined in the creation of a concrete pavilion, where structural rebar is incorporated into the cast, improving stability and enabling more complex designs [16]. This combination of material science and 3D printing allows for greater flexibility in formwork design, though concerns about the environmental impact of plastic waste remain a challenge for the future of these technologies.

In relation to the existing methods, our approach integrates two innovative strategies for sustainable casting. First, we utilize a 3D printed mold that maintains hydration throughout the cast's drying process, ensuring optimal curing conditions and preventing structural issues like cracking. Second, we have selected a biodegradable casting material that can potentially be reused in future casts, contributing to waste reduction and sustainability. While this reuse potential is still being tested, the ability to recycle this material could offer significant environmental benefits in the long term, reinforcing the sustainable nature of wet-mold casting.

3 Methodology

The proposed method for casting individualized concrete components utilizes an uncured, 3D printed mold positioned within an outer container. The set-up consists of the following components as per (Table 1).

The first step is the preparation of an outer container to withstand the hydrostatic pressure of the 3D printed mold, infill, and concrete materials. This consists of a timber or steel box large enough to fit the print form itself, while also accounting for

Table 1 Mold components

Component	Function	Materials
Outer container	Withstand hydrostatic pressures	Timber or steel
Infill	Provide support for 3D printed mold while withstanding pressures of the mold and casting material	3D printed wood infill or similar
Inner 3D printed mold	Dictate shape of the cast via a computer driven robotic arm or gantry	1:1:10 (by weight) mixture of wood chips, methylcellulose, water
Casting material	Takes the desired shape controlled by the 3D printed mold, becoming a continuous custom component when dried or cured	Concrete mix
Reinforcement	Maintain structural integrity throughout the casted unit	Steel rebar

the tool path of the print head and any auxiliary components that may be attached. As such, tall prints necessitating a deep formwork may require additional consideration of toolpath planning.

The 3D printed mold is extruded from a robotic arm or gantry system serving as the core automation system. In this case, a custom mechanical auger-driven extruder was attached to a Kuka KR20-3 robot, enabling precise positional deposition and toolpaths. The extruder compressed and dispensed the organic mixture through a reduced-diameter nozzle, maintaining consistent extrusion pressure and uniform print width throughout the process.

The space between the outer container and the 3D printed mold could be solidly filled with the extruded material. However, to reduce the actual printing time, which would be a main cost driver in industrial production, we opted to separately inject infill material into this space that can be pumped or manually inserted. This material could be the same as the 3D printed material, or a different material of similar consistency. If a different material is preferred, a separating layer could be introduced to prevent contamination of the 3D printing material.

During the printing process itself, the organic material mixture consisting of a 1:1:10 ratio by mass of lightweight, fibrous wood chips, methylcellulose, and water was loaded into a feeder tube to sustain uninterrupted printing (Fig. 2). After 1–3 successive layers of the 3D printed mold, depending on the stability of the form at that time, the infill material was deposited, and concrete was filled into the mold (Fig. 3). The infill material and concrete thereby become the support material for the higher printing layers, enabling the construction of steep inclinations and overhangs.

The synchronized operation of the robotically controlled 3D printed mold and infill material insertion minimized disturbances to the printed material, ensuring the mold retained its intended shape. Once the fill materials were in place, the form became stable and resisted moderate disturbances, such as the deposition of subsequent layers, without deformation.

Fig. 2 Close-up image of the organic material mixture before 3D printing, consisting of a (1:1:10) mixture of fibrous wood chips, methylcellulose, and water

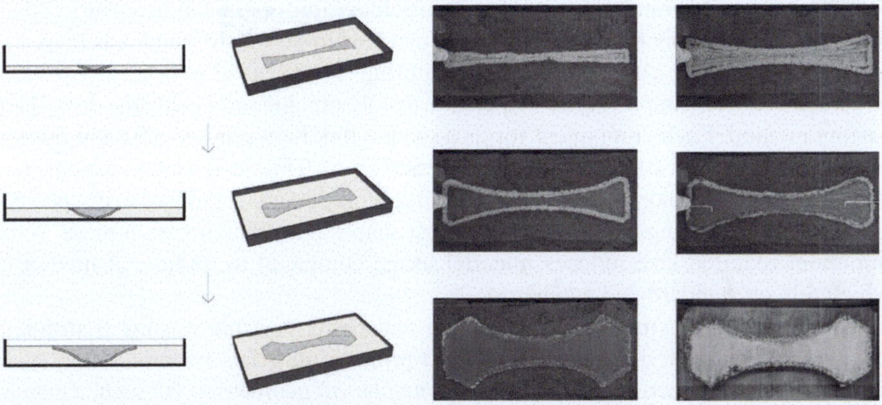

Fig. 3 Representation of the printing process as described, with photos taken during the demonstration print showing the process at various stages

For any elements that were to be placed inside the concrete form, such as steel reinforcement or mounting hardware, this was done at appropriate stages during printing. Elements could cross the print material threshold, provided the organic material could be extruded above or around them.

Following the completion of the print, the entire volume is left to cure for a period sufficient for the concrete to develop strength for handling. After this curing period, the form is removed from the infill and containment area. The printed material, which remained slightly drier than in its pre-printing state, was scraped off and stored for reuse. Residual material attached to the concrete form was removed, leaving the final concrete component to fully cure.

4 Results and Discussion

4.1 Case Study

The proposed method has various potential applications, within and beyond the context of structural and architectural prefabricated building solutions. As a demonstration of the method to show one of these possible solutions and to test the feasibility of the method itself, the wet-mold casting was applied to the design of a building case study and the production of one of its structural components.

The design consists of a tower portion, a support podium, and a free-form semi-structural exoskeleton that wraps around both (Fig. 4). The free-form design is optimized for structural performance, minimizing the required concrete volume by tailoring each component to the specific structural demands at its location. This not only reduces material waste but also lowers the building's CO_2 (carbon-dioxide) footprint. Gravitational simulations were used to achieve this optimization, ensuring that the structural components of the skin were oriented to best distribute loads (Fig. 5).

The skin itself is a modular design, featuring a core metal lattice system with structural concrete supporting it, shaped into a more organic form using the described casting method. Each component forms a unique link between two adjacent nodes, which connects up to six separate components (Fig. 6). This design approach ensures that each of the approximately 13,000 individual components is tailored to its specific role, possibly resulting in 13,000 unique cast shapes. Customizing each piece to its minimum required size reduces material usage compared to traditional methods, which rely on standardized geometries.

A design like this might be challenging and costly to achieve using traditional casting techniques. Using non-reusable 3D printing materials for molds may also lead to increased material waste when creating one-off geometries. As such, a manufacturing technology consisting of flexible 3D printing and easily reusable materials is beneficial for reducing cost while maintaining design integrity.

Currently, architectural detailing and shaping are often viewed as constrained by the inflexible manufacturing capabilities of traditional methods. While complex forms in concrete can be achieved, these methods are typically resource-intensive and costly. By contrast, the described process facilitates the efficient production of highly customized forms, resulting in dramatic cost reductions and a streamlined workflow for creating each unique component. The modular exoskeleton design illustrates how tailored components may minimize material usage, reduce environmental impact, and maintain structural integrity, potentially accommodating more complex and innovative architectural designs.

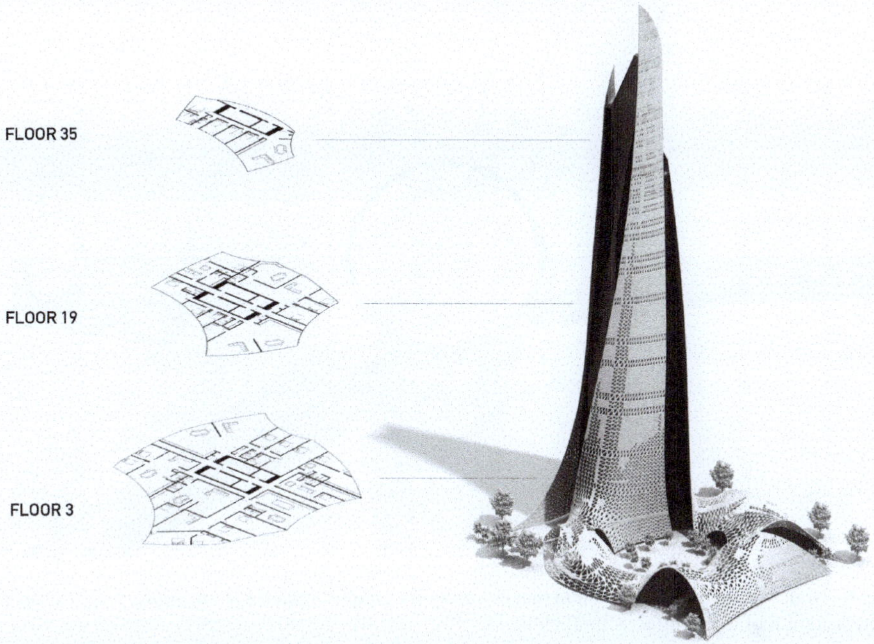

Fig. 4 Showcase of the case study tower, with several floors pulled out to highlight the shifting, organic nature of the design

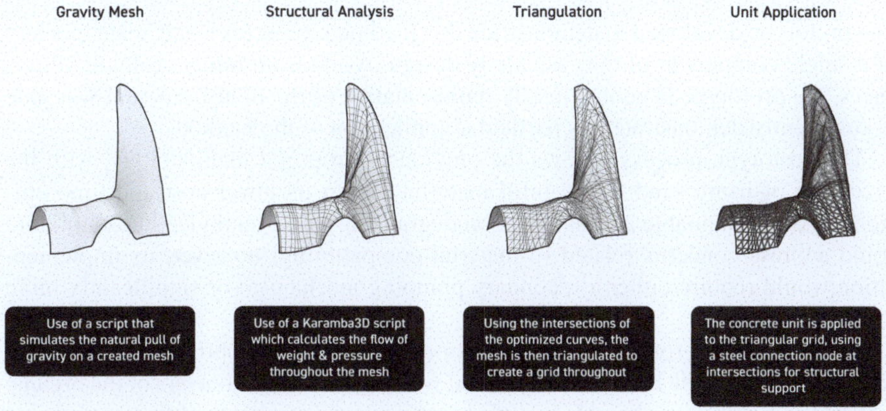

Fig. 5 Representation of scripts used to calculate the distribution of forces on a gravitationally simulated mesh

Fig. 6 Node within the lattice system that connects six unique elements at one point, with a simple representation of a joining structural element

4.2 Prototype Unit

The prototype developed to demonstrate this method represents a half scale mockup of a single component of the case study design. Aside from minor aesthetic differences, the prototype is geometrically representative of the final form, and was able to sufficiently demonstrate this particular application of the method.

The prototype process follows the previously proposed methodology, with the exception of using sand as the infill material due to its lower cost and time efficiency. As a sustainable alternative, a biodegradable wood-methylcellulose mixture could address concerns related to material compatibility; however, its implementation would require either a secondary pumping mechanism or significantly more time.

The final printed form was set as a component measuring approximately 863 mm long by 257 mm wide by 54 mm thick, and was printed using 11 layers of the organic material at 5 mm layer height. Included was a simple steel embed that served as a placeholder for a reinforcement system. The placeholder connection points were embedded within, and projected beyond the form, and served to demonstrate the ability for secondary elements to be placed within the printed mold.

The finish surface of the unit, owing to the unique printing process of the mold, resulted in inverted layer lines appearing on the surface of the concrete unit (Fig. 7), and the texture of the organic, fibrous printing material was imprinted upon that. Due

Fig. 7 Close-up view of the final cast surface, highlighting the finish and the layer lines that appeared

to the process of scraping off the residual wood pieces with brushes, a number of wood pieces remained embedded in the final unit.

4.2.1 Accuracy and Dimensional Stability

The final print demonstrated a high level of dimensional accuracy. The overall length of the piece was designed to be 863 mm, while the final component had a length of 857 mm. The width at the widest and thinnest sections similarly deviated by a few millimeters from the target geometry, 248 mm instead of 257 mm at the widest point. The overall height of the piece was 51 mm instead of the intended 54 mm. In the vertical direction, it is possible that the hydrostatic pressure in combination with cavities from the printing process may have had an impact. This should be investigated further. Also, the measurements are imprecise due to the final texture of the piece, and were taken at locations where the texture on the surface almost entirely gives way to solid concrete, eliminating the above-described texture from these measurements. The texture projects by up to 8–10 mm, with longer projections easily broken off and removed.

4.3 Challenges and Potential Solutions

4.3.1 Infill Materials and Structural Considerations

As per the literature review, the 3D printing of biodegradable materials, especially those based on water as the solvent, often faces challenges of the stability of the print in its wet state. Steep wall inclinations and overhangs are difficult to achieve. We overcome those challenges with our proposed method as the print bed is fully filled, beyond the printed material, with the casting material and the infill, which together form the support structure for the 3D printed mold. Any wall inclination and overhangs are therefore easily achievable with our method.

A concern of the infill material is its compressibility, including air pockets that might occur during its insertion. In the production of the prototype, sand was used as the infill material. It offered a more cost-effective option for making a single mold, as opposed to also using the print material as infill, and did not compress under the hydrostatic pressure of the concrete. Wet concrete has a substantial weight, approximately 2400 kg/m3. Wet sand is similarly heavy compared to the concrete and is less prone to deformations compared to the print material. By having it push against the concrete, with the print shell in between, it is these properties that serve to hold the material in place and give this material the rigidity needed to maintain its geometry. Using the print material as the infill may suffice, especially on smaller prints. However, further testing would be needed to evaluate the accuracy and the true extent of the deformation of the material under the hydrostatic pressure of the concrete. If the results are within acceptable levels, it will be beneficial to avoid a separate infill material, as for example the reuse of the material would be simpler, avoiding the need to remove any sand.

4.3.2 Composition and Challenges of the Print Material

There are several other potential improvements of the proposed method that should be the target of future development. First, while a successful composition for this proof-of-concept of wet-mold casting, the print material itself should be optimized. The current composition is a mixture of water, methylcellulose, and wood chips. Due to variations in fiber dimensions, the overall consistency of the mixture is not completely consistent. The size and texture of the wood chips have the potential to be improved by using smaller fibers or a different biodegradable or recyclable material altogether. Additionally, other agents to tune the viscosity could be explored, as well as potential additives to reduce the chance of the organic material growing mold.

Another exploration is to verify the reusability of the print material. In each print, some material will be lost in its removal from the outer container and the cast component, while a certain amount of material contamination is expected, especially if a different material is used for the infill. In a repeated recycling of the material, the contamination will eventually render the material unusable for further prints and require its biodegradation. Additionally, the lost material will need to be replaced. While we are certain that we can achieve a high degree of material reuse, precise data would be of interest.

4.3.3 Surface Quality and Texture

The finish surface of the final component is unique in its appearance, owing to the composition of the printing material. It has a rough texture pockmarked with places where the concrete began to penetrate the material, on top of the appearance of the inverse of layer lines found in traditional 3D printing (Fig. 8). While from an aesthetic standpoint, this can be considered unique and give parts printed using this technique a particular character not achieved by other methods, it could be considered undesirable by other designers. It is worth pursuing different printing material compositions and printing techniques to allow for the potential to refine and customize the surface finish.

We also encountered several wood fibers becoming embedded in the concrete itself. If this is undesirable, a slurry composition with smaller particles could be developed, or there could be procedures for their removal, possibly by burning or

Fig. 8 Close-up view showing the final surface finish as well as the residual wood fibers from the print material still embedded in the concrete cast

dissolving the fibers. Further exploration is necessary for the integration of non-printed hardware into the final form. While it was shown to be feasible during experimentation, the parts used to test this were relatively simple and not representative of many potential applications.

4.4 Comparison to Similar Methods

As per the literature review, other professionals and scholars have used 3D printed molds for concrete casting, including with 3D printed molds made from biodegradable materials. However, in all cases, the mold is left to solidify or cure before being filled with concrete. By comparison, our process significantly speeds up not only the time it takes to cast an element, on account to leaving the mold wet, but the containment area and the infill also significantly improve the print's stability against the concrete's hydrostatic pressure. On the other hand, Burger [16] outlines a process that, similar to the process described here, casts simultaneously with the printing process and thereby increases efficiency. However, due to the mold's plastic nature, the method does not use a material that can be easily and immediately reused. It does negate the need for a containment area and an infill material, however. A further point of similarity is that Burger describes a method for encasing structural steel components in the cast element in a way that is similar to the method proposed here.

5 Conclusion

We presented a novel method for the 3D printed wet-mold casting of concrete. The technical feasibility of the method has been shown, demonstrating a level of structural stability in the print that allows for economically produced one-off geometries. In contrast to other methods, which require the mold's stability during the printing process, our proposal allows for any desired form, including inclinations and overhangs. In addition, the degree of accuracy in the final cast is close to the initial target geometry.

This research has demonstrated a novel approach to integrating 3D printing technology with traditional concrete casting, offering a sustainable and efficient solution to challenges in contemporary construction. The slurry material developed for this process effectively enables the production of highly customized molds, which not only support intricate architectural designs but also address limitations in flexibility, material efficiency, and environmental impact, aligning with broader industry goals of reducing waste and promoting eco-friendly practices.

Key advancements include the ability to integrate traditional reinforcement techniques, such as steel rebar and structural connectors, into the casting process without compromising the design. These capabilities overcome the structural limitations of directly 3D printed concrete and enable the creation of complex forms that are both elegant and functional. The case study highlights the method's transformative potential by producing unique design components optimized for structural performance and material efficiency, in line with principles in parametric modeling and topology optimization.

Looking ahead, several areas of refinement and exploration could enhance the method's viability for large-scale implementation. Investigating alternative slurry compositions, improving slurry reusability, and optimizing extrusion techniques would further increase the efficiency and scalability of the process. Automation of the infill placement and exploration of the mold removal could increase consistency and reduce production errors. Additionally, understanding the boundaries of this technology, such as minimum viable mold sizes and the integration of prefabricated components, will provide a clearer framework for its application in various construction scenarios.

This method offers progress in merging the flexibility of 3D printing with the structural robustness of traditional concrete. By exploring challenges related to material waste, sustainability, and complex design, this approach will contribute to a more innovative, efficient, and sustainable future in construction. With continued development, it has the potential to reshape how architectural components are conceived, manufactured, and implemented.

References

1. Schutter, G.D., Lesage, K., Mechtcherine, V., Nerella, V.N., Habert, G., Agusti-Juan, I.: Vision of 3D printing with concrete—technical, economic and environmental potentials. Cem. Concr. Res. **112**, 25–36 (2018). https://doi.org/10.1016/j.cemconres.2018.06.001
2. Panda, B., Tan, M.: Experimental study on mix proportion and fresh properties of fly ash based geopolymer for 3D concrete printing. Ceram. Int. **44**(9), 10258–10265 (2018). https://doi.org/10.1016/j.ceramint.2018.03.031
3. Bhattacherjee, S., Basavaraj, A.S., Rahul, A.V., Santhanam, M., Gettu, R., Panda, B., Schlangen, E., Chen, Y., Copuroglu, O., Ma, G., Wang, L., Benign, M.A.B., Mechtcherine, V.: Sustainable materials for 3D concrete printing. Cem. Concr. Compos. **122**, 104156 (2021). https://doi.org/10.1016/j.cemconcomp.2021.104156
4. Sahin, H., Mardani, A., Beytekin, H.E.: Effect of silica fume utilization on structural build-up, mechanical and dimensional stability performance of fiber-reinforced 3D printable concrete. Polymers **16**(4), 556 (2024). https://doi.org/10.3390/polym16040556
5. Kaszynska, M., Skibicki, S., Hoffmann, M.: 3D concrete printing for sustainable construction. Energies **13**(23), 6351 (2020). https://doi.org/10.3390/en13236351
6. Lloret-Fritschi, E., Wangler, T., Gebhard, L., Matta-Falcon, J., Mantellato, S., Scotta, F., Burger, J., Szabo, A., Ruffray, N., Reiter, L., Boscaro, F., Kaufmann, W., Kohler, M., Gramazio, F., Flatt, R.: From smart dynamic casting to a growing family of digital casting systems. Cem. Concr. Res. **134**, 106071 (2020). https://doi.org/10.1016/j.cemconres.2020.106071
7. Vukorep, I., Zimmermann, G., Sablotny, T.: Robot-controlled fabrication of sprayed concrete elements as a cyber-physical-system. In: Bos, F.P., Lucas, S.S., Wolfs, R.J.M., Salet, T.A.M. (eds.) Second RILEM International Conference on Concrete and Digital Fabrication, Eindhoven, July 2020. RILEM, vol. 28, pp. 967–977. Springer, Heidelberg (2020)
8. Jipa, A., Bernhard, M., Meibodi, M., Dillenburger, B.: 3D printed stay-in-place formwork for topologically optimized concrete slabs. In: Kory, B. (ed.) 2016 TxA Emerging Design +

Technology Conference, San Antonio, November 2016. TxA Emerging Design + Technology Conference Proceedings, vol. 77, pp. 97–107. Texas Society of Architects, Austin (2016)
9. Jipa, A., Meibodi, M.A., Giesecke, R., Shammas, D., Leschok, M., Dillenburger, B.: 3D printed formwork for prefabricated concrete slabs. Paper presented at the 3rd 1st International conference on 3D construction printing, Melbourne, November 2018. International Conference on 3D Construction Printing
10. Yang, W., Wang, L., Ma, G., Feng, P.: An integrated method of topological optimization and path design for 3D concrete printing. Eng. Struct. **291**, 116425 (2023). https://doi.org/10.1016/j.engstruct.2023.116435
11. Garcia-Alvarado, R., Moroni-Orellana, G., Banda-Perez, P.: Architectural evaluation of 3D printed buildings. Buildings **11**(6), 254 (2021). https://doi.org/10.3390/buildings11060254
12. Sakin, M., Kiroglu, Y.: 3D printing of buildings: construction of the sustainable houses of the future by BIM. Energy Procedia **134**, 702–711 (2017). https://doi.org/10.1016/j.egypro.2017.09.562
13. Leschok, M., Cheibas, I., Piccioni, V., Seshadri, B., Schluter, A., Gramazio, F., Kohler, M., Dillenburger, B.: 3D printing facades: Design, fabrication and assessment methods. Autom. Constr. **152**, 104918 (2023). https://doi.org/10.1016/j.autcon.2023.104918
14. Piccioni, V., Lechok, M., Grobe, L.O., Wasilewski, S., Seshadri, B., Hiscier, I., Schluter, A.: Tuning the solar performance of building facades through polymer 3D printing: toward bespoke thermo-optical properties. Adv. Mater. Technol. **8**(6), 2201200 (2023). https://doi.org/10.1002/admt.202201200
15. Khan, M., Varadharajan, T., Keller, Z., Meibodi, M.: The robotic 3D printed biodegradable wood-based formwork for cast-in-place concrete structures. In: Crawford, A., Diniz, N., Beckett, R., Vanucchi, J., Swackhamer, M. (eds.) Habits of the anthropocene: scarcity and abundance in a post-material economy, Denver, October 2023. Proceedings of the 43rd Annual Conference of the Association for Computer Aided Design in Architecture, vol. 22, pp. 174–183. Acadia, Denver. (2023)
16. Burger, J., Aejmelaeus-Lindstrom, P., Gurel, S., Niketic, F., Lloret-Fritschi, E., Flatt, R.J., Gramazio, F., Kohler, M.: Eggshell Pavilion: a reinforced concrete structure fabricated using robotically 3D printed formwork. Constr. Robot. **7**, 213–233 (2023). https://doi.org/10.1007/s41693-023-00090-x
17. Bernhard, M., Hansmeyer, M., Dillenburger, B.: Volumetric modelling for 3D printed architecture. In: Hesselgren, L., Kilian, A., Malek, S., Olsson, K.-G., Sorkine-Hornung, O., Williams, C. (eds.) Advances in Architectural Geometry 2018, Vienna, September 2018. De Gruyter STEM, vol. 6, pp. 392–415. Klein, Berlin (2018)
18. Ko, C.H.: Constraints and limitations of concrete 3D printing in architecture. JEDT **20**(5), 1334–1348 (2021). https://doi.org/10.1108/JEDT-11-2020-0456
19. Khan, S., Koc, M., Al-Ghamdi, S.: Sustainability assessment, potentials and challenges of 3D printed concrete structures: a systematic review for built environmental applications. J. Clean. Prod. **303**, 127027 (2021). https://doi.org/10.1016/j.jclepro.2021.127027
20. Abdalla, H., Fattah, K., Abdallah, M., Tamimi, A.K.: Environmental footprint and economics of a full-scale 3D printed house. Sustainability **13**(21), 11978 (2021). https://doi.org/10.3390/su132111978
21. Zhao, Z., Ji, C., Xiao, J., Yao, L., Lin, C., Ding, T., Ye, T.: A critical review on reducing the environmental impact of 3D printing concrete: material preparation, construction process and structure level. Constr. Build. Mater. **409**, 133887 (2023). https://doi.org/10.1016/j.conbuildmat.2023.133887
22. Salandin, A., Quintana-Gallardo, A., Gomez-Lozano, V., Guillen-Guillamon, I.: The first 3D printed building in Spain: a study in its acoustic, thermal, and environmental performance. Sustainability **14**(20), 13204 (2022). https://doi.org/10.3390/su142013204
23. Beyhan, F., Selcuk, S.: 3D printing in architecture: one step closer to a sustainable built environment. In: Beyhan, F., Selcuk, S.A. (eds.) Proceedings of 3rd International Sustainable Buildings Symposium (ISBS 2017), Dubai, March 2017. Lecture Notes in Civil Engineering, vol. 6, pp. 253–268. Springer, Cham. (2018)

24. Saidin, W., Wagiman, A., Ibrahim, M.: Development of wood-based composites material for 3D printing process. AMM **315**, 987–991 (2013). https://doi.org/10.4028/www.scientific.net/AMM.315.987
25. Kong, M., Qin, Z., Zhang, P., Xie, G., Wang, H., Wang, J., Guan, F., Yang, W., Qiu, Z.: Study on modified poplar wood powder/polylactic acid high toughness green 3D printing composites. Int. J. Biol. Macromol. **228**, 311–322 (2023). https://doi.org/10.1016/j.ijbiomac.2022.12.197
26. Bierach, C., Coelho, A.A., Turrin, M., Asut, S., Knaack, U.: Wood-based 3D printing: potential and limitation to 3D print building elements with cellulose & lignin. Archit. Struct. Constr. **3**, 157–170 (2023). https://doi.org/10.1007/s44150-023-00088-7
27. Thakur, S., Sci, C., Kearney, L.T., Saadi, M.A.S.R., Meyer, M.D., Naskar, A.K., Ajayan, P.M., Rahman, M.M.: Three-dimensional printing of wood. Sci. Adv. **10**(11), 3250 (2024). https://doi.org/10.1126/sciadv.adk3250
28. Henke, K., Treml, S.: Wood based bulk material in 3D printing processes for applications in construction. Eur. J. Wood Wood Prod. **71**, 139–141 (2012). https://doi.org/10.1007/s00107-012-0658-z
29. Klemmt, C., Aghaei Meibodi, M., Beaucage, G., McGee, W.: Large-scale robotic 3D printing of plant fibre and bioplastic composites. In: eCAADe Proceedings (2022). https://doi.org/10.52842/conf.ecaade.2022.1.009
30. Rosenthal, M., Henneberger, C., Gutkes, A., Bues, C.-T.: Liquid deposition modeling: a promising approach for 3D printing of wood. Eur. J. Wood Wood Prod. **76**, 797–799 (2017). https://doi.org/10.1007/s00107-017-1274-8
31. Pringle, A.M., Rudnicki, M., Pearce, J.M.: Wood furniture waste-based recycled 3-D printing filament. For. Prod. J. **68**(1), 86–95 (2018). https://doi.org/10.13073/FPJ-D-17-00042
32. Kam, D., Layani, M., BarkaiMinerbi, S., Orbaum, D., BenHarush, S.A., Shoseyov, O., Magdassi, S.: Additive manufacturing of 3D structures composed of wood materials. Adv. Mater. Technol. **4**(9), 1900158 (2019). https://doi.org/10.1002/admt.201900158
33. Ayrilmis, N.: Effect of layer thickness on surface properties of 3D printed materials produced from wood flour/PLA filament. Polym. Test. **71**, 163–166 (2018). https://doi.org/10.1016/j.polymertesting.2018.09.009
34. Yang, T.-C., Yeh, C.-H.: Morphology and mechanical properties of 3D printed wood fiber/polylactic acid composite parts using fused deposition modeling (FDM): the effects of printing speed. Polymers **12**(6), 1334 (2020). https://doi.org/10.3390/polym12061334
35. Wimmer, R., Steyrer, B., Woess, J., Koddenberg, T., Mundigler, N.: 3D printing and wood. In: Gurau, L., Campean, M., Ispas, M. (eds.) Proceedings of the International Conference "Wood Science and Engineering in the Third Millennium", Brasov. Applied Sciences, vol. 10, pp 145–150. Transilvania University of Brasov, Brasov (2015)
36. Mayo, S.C., Chen, F., Evans, R.: Micron-scale 3D imaging of wood and plant microstructure using high-resolution X-ray phase-contrast microtomography. J. Struct. Biol. **171**(2), 182–188 (2010). https://doi.org/10.1016/j.jsb.2010.04.001
37. Burger, J., Lloret-Fritschi, E., Scotto, F., Demoulin, T., Gebhard, L., Mata-Falcon, J., Gramazio, F., Kohler, M., Flatt, R.J.: Eggshell: ultra-thin three-dimensional printed formwork for concrete structures. 3D Print. Addit. Manuf. **7**(2), 48–59 (2020). https://doi.org/10.1089/3dp.2019.019
38. Mozaffari, S., Bruce, M., Clune, G., Xie, R., McGee, W., Adel, A.: Digital design and fabrication of clay formwork for concrete casting. Autom. Constr. **154**, 104969 (2023). https://doi.org/10.1016/j.autcon.2023.104969
39. Mora, S., Pugno, N.M., Misseroni, D.: 3D printed architected lattice structures by material jetting. Mater. Today **59**, 107–132 (2022). https://doi.org/10.1016/j.mattod.2022.05.008
40. Bedarf, P., Dutto, A., Zanini, M., Dillenburger, B.: Foam 3D printing for construction: a review of application, materials, and processes. Autom. Constr. **130**, 103861 (2021). https://doi.org/10.1016/j.autcon.2021.103861
41. Jipa, A., Dillenburger, B.: 33D printed formwork for concrete: state-of-the-art, opportunities, challenges, and applications. 3D Print. Addit. Manuf. **9**(2), 84–107 (2022). https://doi.org/10.1089/3dp.2021.0024

42. Curth, A., Pearl, N., Wissemann, E., Cousin, T., Alkhayat, L., Jackow, V., Lee, K., Moldow, O., Ismail, M., Mueller, C., Sass, L.: EarthWorks: zero waste 3D printed earthen formwork for shape-optimized, reinforced concrete construction. Constr. Build. Mater. **449**, 138387 (2024). https://doi.org/10.1016/j.conbuildmat.2024.138387
43. Leschok, M., Dillenburger, B.: Sustainable thin-shell 3D printed formwork for concrete. In: Gengnagel, C., Baverel, O., Burry, J., Thomsen, M.R., Weinzierl, S. (eds.) Impact: Design With All Senses—Proceedings of the Design Modelling Symposium, Berlin. DMS: Design Modelling Symposium Berlin, vol. 1, pp. 487–501. Springer, Cham (2019)

Brendan Kiser received his B.S. in Architecture from the University of Cincinnati in 2023. Since graduating Brendan has worked at RSL Architecture, contributing to the design and execution of themed and commercial projects for multiple Six Flags theme parks including Cedar Point and Schlitterbahn. Alongside his full-time role, Brendan collaborates with Architect Karl Hoalst on residential renovations, further developing his skills in practical design development and construction documentation. He is currently pursuing a Master of Architecture through Southern Illinois University, with the goal of becoming a licensed architect following graduation.

Ryan Kennedy received his B.S. in Architecture from the University of Cincinnati in 2023. During his studies, and following his graduation, he worked at Champlin/EOP Architects in Cincinnati, Ohio, as a member of the design team. He plays a significant role in an initiative to improve design quality, as well as technological proficiency in the firm, helping elevate the firm from a purely service oriented office into an award winning design firm. This includes specifically building up and managing both architectural visualization capabilities, as well as a specialty in parametric design integration. Ryan is currently pursuing licensure as an architect, continuing his ongoing studies.

Lucas Palicki received his B.S. in Architecture from the University of Cincinnati in 2023. As part of his studies, Lucas participated in multiple internships with award-winning firms. His study of parametric architecture during his capstone project inspired his participation in this publication. While studying at UC, Lucas maintained his lifelong passion for golf, competing both individually and on the school's club golf team. Through this process, he recognized his professional calling: to integrate his architectural skills with his passion for golf. Lucas now operates his own golf consulting business, Palicki Golf Design. With this experience, Lucas will pursue licensure with the American Society of Golf Course Architects.

Chapter 10
3D Printing of Wood and Bioplastic Composites for Architectural Applications

Christoph Klemmt and Gregory Beaucage

Abstract Various projects have used additive manufacturing for architectural applications, with uses ranging from the production of structural joints to molds for concrete casting, façade elements, prefabricated components and in-situ printed structures. However, many of those methods require high energy consumption, use materials that have high CO_2 emissions or that are not renewable. This project therefore investigated the use of wood and biodegradable binders for large-scale 3D printing. Different fibers and binders were tested for their extrudability, physical properties, and drying behavior, especially regarding the reduction of shrinkage. Relationships are described for fiber and binder properties to shrinkage and mechanical performance. Suitable material compositions were used for the construction of items ranging in scale from product design to a large architectural column. The lightweight materials, akin to wood foams, are deemed suitable for various architectural applications. While the materials are strong enough for structural applications, further work is required to improve the materials' safety and durability.

Keywords Architecture · 3D printing · Wood · Plant fiber · Additive manufacturing

1 Introduction

The construction industry has been recognized as one of the main contributors to CO_2 emissions and global warming, both through the use of energy during the construction of buildings, as well as through emissions caused by their use during the lifespan of the completed building. Furthermore, the demolition of buildings is a main contributor to waste, with limited options for recycling that are usually only available for select

C. Klemmt (✉)
College of Design, Architecture, Art and Planning, University of Cincinnati, Cincinnati, OH, USA
e-mail: christoph@orproject.com

G. Beaucage
Chemical and Materials Engineering, University of Cincinnati, Cincinnati, OH, USA

© The Author(s), under exclusive license to Springer Nature Singapore Pte Ltd. 2026

C. Klemmt and G. Beaucage (eds.), *Biodegradable 3D-Printing*, The Materials Research Society Series, https://doi.org/10.1007/978-981-95-4025-9_10

Fig. 1 3D printed prototype in the form of an architectural column of 2 m height

building parts such as the building's metal components. It is therefore imperative to reconsider the materials and methods that we use to construct buildings, with a focus on renewable resources, low-energy methods, and reusable, recyclable or biodegradable materials (Fig. 1).

Among novel construction methods, 3D printing (3DP) is gaining in relevance in the construction industry. Available methods include the 3D printing of concrete, polymers, metal and soil. Contrary to subtractive manufacturing methods, additive manufacturing (AM) techniques in general offer the sustainable benefit of only using the material needed for the final piece. However, there are also various drawbacks to some of the methods currently employed. One main concern is energy usage: in the case of concrete 3DP in the production of the cement, or in the case of 3DP of metals or thermoplastics in the required heating during the actual AM process. A further concern is the construction waste of those materials that cannot or are unlikely to be recycled or reused. This often includes polymers, but also seemingly environmentally friendly materials such as soil if it has been enhanced with additives such as lime.

To address some of those concerns, we therefore hypothesize that the 3D printing of plant fibers with biodegradable binders is a suitable alternative for various applications. In 2-dimensions, established industries as well as recycling streams exist that produce paper, cardboard or timber boards. With the possibility to expand those industries to produce 3-dimensional products, we therefore present a method for the large-scale 3D printing of wood composites. The biodegradable material consists of wood shavings and a biodegradable, cellulose-based binder. The material is therefore fully derived from wood, with a high wood content above 60%. While prone to hydrolysis and fire and therefore not suitable for permanent structural applications without additional treatment, the material exhibits remarkable strength in both compression and tension. It therefore has wide-ranging possible applications including interior cladding panels, stage sets, furniture, or molds for casting. A prototype architectural column has been constructed using this method (1). The material comes either from renewable sources or from the existing and extensive timber waste streams. At the end of its life cycle, it can be either reused, recycled, or biodegraded.

2 Related Work

2.1 3D Printing Architecture

While the 3D printing of small-scale objects has been available for several decades, commercially viable 3D printing at the scale of architecture and of buildings or building components is still relatively new. Direct 3DP, as well as the 3DP of molds, has been used to produce a variety of building components at different scales. Directly printed objects include load-bearing components such as walls, columns or slabs, complex metal joints, or cladding panels [1]. Molds have been 3D printed especially for the casting of concrete, either as load-bearing components [2, 3] or as cladding panels [4].

Early work in the 3D printing of concrete was carried out by Joseph Pegna [5], while Behrokh Khoshnevis developed "contour crafting," a method of building up structures layer by layer from concrete [6]. The fast-curing concrete paste is extruded from a nozzle. A method of large-scale binder jetting was developed by Dini that uses a sandstone material and has led to the construction of a sculpture of 3 m x 3 m x 3 m in 2009 [7].

The most prevalent commercially available method at the moment is the large-scale 3D printing of concrete. Notable was the development of concrete 3D printing by the company Winsun, which constructed several houses in 2013 [8]. The commercial 3D printing of concrete has since been taken on by various companies, including Peri [9] and Icon [10]. Research addresses issues such as the integration of reinforcement [11] or the utilization of the 3DP concrete as lost formwork [12, 13].

Various non-cementitious methods and materials have been developed for large-scale 3D printing. Some of these include thermoplastics [1, 14], metals [15], spray foam [16], and soil and lime composites [1].

2.2 3D Printing of Biodegradable Materials

While AM offers ecologic possibilities by only utilizing the required material and thereby reducing wastage during the manufacturing process, many of the materials and techniques used at the large-scale have significant drawbacks in regards to their sustainability: The 3D printing of metal and thermoplastics uses large amounts of energy in their production [1, 14, 15], plastics and concrete are usually not biodegradable [1, 8], and especially cementitious materials have large CO_2 emissions during their production [8, 9].

AM materials with more sustainable possibilities include those that exist abundantly in nature such as unfired soil and silicates, and biodegradable materials. The 3D printing of biodegradable materials is commonly based on plant fibers with a biodegradable binder such as xanthan gum, cellulose ethers, lignin or chitin, and often water as a solvent. This results in a reversible material that dries through the evaporation of the solvent but can be reused upon rehydration. Various AM methods have been developed for those materials, ranging from the small scale to the size of architectural installations.

A commonly stated aim is the maximization of wood content. At the small scale, Kariz et al. proposed a method using polyvinyl acetate and urea formaldehyde, with a limited wood content of 12–25% [17]. However, while Rosenthal et al. and Kam et al. managed to increase the wood content to 90%, they observed shrinkage during drying of around 20% [18, 19]. The composition of Rosenthal is similar to the formulations proposed in this chapter, however with different fiber dimensions. Hausmann et al. developed a method whereby after the printing process, the solvent is exchanged prior to drying the object. The objects retain their shape but shrink by 30% [20]. Deville et al. proposed the freezing of the 3D printed objects, which can be used for a high control over the fiber orientation in the objects [21]. Jiang et al. used the cross-linking of lignin to produce small-scale objects, with the removal of water by means of freeze-drying [22]. Thakur et al. similarly used a combination of cellulose fibers with lignin as the binder [23].

At the large scale, the 3DP of biodegradable materials has found interest in the disciplines of art and architecture. Goidea et al. utilized mycelium as the binding agent. A slurry comprising a growth medium and spores is extruded, with the fungus growing to stabilize the extrudate [24]. Caracassi et al. developed a series of material compositions of clay and soil with added plant fibers [25]. Dritsas et al. used the co-extrusion of cellulose and chitin to construct a variety of larger objects that are fully biodegradable [26]. Rossi et al. extruded objects composed of cellulose and other plant fibers with xanthan gum as the binder [27].

Across different scales and using different binders, the shrinkage of the materials during the drying process has been a consistent problem. Unlike cementitious materials, where the evaporation of the solvent only causes small and usually acceptable shrinkage, the deformations of the biodegradable materials have been documented in the range of 20% or more. Proposed solutions such as the solvent-exchange or freeze-drying may only be feasible at the small scale [20, 21]. Other proposals to overcome this drawback have been design adjustments to accommodate for the anticipated deformations [27, 28].

In order to move toward more sustainability in 3D printing for construction, it is imperative to develop methods that are large-scale, low-cost, low-energy, and use renewable and biodegradable materials. We therefore present a method for the large-scale 3D printing of wood and bioplastic composites.

3 Method

3.1 Printing Mechanism

The 3D printing of the pieces of this research was achieved by means of a paste extruder. The custom-built extruder is attached to an industrial robotic arm that facilitates the extruder's movement in 3D-space, based on a programmed toolpath. For this project, a Kuka KR20 was used, which has a reach of approximately 1.6 m.

The slurry is either pumped or manually filled into the auger-actuated extruder. The extruder has a diameter of 100 mm, while nozzles were used with diameters in the range of 15 mm–21 mm. The motor has a torque of 20.7Nm. At a slurry viscosity of 500,000cps, we estimate a head pressure in the range of 0.5 MPa, and a flow rate in the range of 16 mL/s.

3.2 Material Compositions

We have tested various types of fibers, binders, solvents, and additives for this study. The main criteria in selecting suitable compositions were the final compressive and tensile strength, cost, and ease of production. Biodegradability or a benign environmental impact were preconditions for consideration of the ingredients.

3.2.1 Fibers

All the fibers investigated were renewable and biodegradable plant fibers. Those included wood shavings, saw dust, refined cellulose, flax, grass fibers, rice husks, corn husks, coconut fibers, and hemp. Some of those were deemed unsuitable based on initial tests prior to a systematic evaluation. A further investigation of those materials was therefore abandoned but may yield successful results with relevant adjustments.

Unsuitability of those fibers was especially caused for reasons of processability and structural integrity. Coconut fibers, hemp and softwood strips with lengths >50 mm were unsuitable for the specific extrusion system due to processing difficulties such as a clogging of the extruder or nozzle. This was mostly caused by a large fiber length or high stiffness of the individual fibers. Those caused a clogging of the extrusion mechanism or difficulties in maintaining the shape of the printed bead after extrusion. A utilization of those fibers at different dimensions or with different processing parameters might lead to successful results.

Saw dust, rice husks, and corn husks had insufficient structural integrity in the final piece. Soft wood in the form of saw dust loses the strength of the individual fibers so that the material only acts as a filler, while the binder must provide the required tensile strength.

On the contrary, the longer fibers of both unrefined softwood shavings as well as refined cellulose (mixed softwood and hard wood) provided tensile resistance, leading to a significantly better tensile performance of the samples. While the Kraft-refined cellulose fibers offered good results and led to successful 3D prints, this study concentrated on pine shavings, with fiber dimensions in the range of width 1–5 mm, length 5–20 mm, and thickness 0.01–1 mm.

In order to overcome processability issues due to the stiffness of the softwood fibers, those were soaked in the solvent prior to extrusion, which led to a significant increase in their bending behavior. This ensured a successful movement of the slurry through the extrusion mechanism.

3.2.2 Binders

The binders investigated included starch, agar, xanthan gum, as well as cellulose ethers such as methylcellulose or hydroxypropyl methylcellulose. Lime was tested as a binder with hemp fibers, however not further pursued for this study due to the above-mentioned difficulties of processing the hemp fibers. Especially the edible binders such as starch and agar were more susceptible to biological decay. Cellulose ethers were found to produce high strength materials. Furthermore, different types of cellulose ethers are commercially available at various viscosities. This allowed for a tuning of the slurry viscosity without the need to alter its solvent content and thereby the material's drying behavior, or the binder to fiber ratio of the final product.

The cellulose ethers are biodegradable, ensure a viscosity of the slurry that allows for an extrusion-based 3D printing, and provide a strong bond between the plant fibers. However, due to a lack of chemical curing, the bond is reversible upon water exposure and the composite material is susceptible to hydrolysis unless otherwise treated post extrusion.

3.2.3 Solvents

Water was used as the solvent. Further experiments with ethanol, using cellulose fibers, resulted in successful 3D prints at the small scale. The use of ethanol resulted in a faster evaporation of the solvent, however, the cost and ease of use made water the preferred option.

4 Material Properties

4.1 Fiber-Swell

Due to the high percentage of solvent in most biodegradable 3DP slurries, the extruded material contracts significantly upon evaporation of the solvents after extrusion. This has been identified as a problem by different researchers working on similar biodegradable materials for 3DP. In the literature, different researchers responded to this by calculating the shrinkage and adjusting the tool paths accordingly [28, 27].

Instead, we identified a possibility to counteract the contraction by causing an expansion of the slurry upon exiting from the nozzle, caused by inducing a swell beyond the already occurring dye-swell. We were able to achieve this by utilizing the plant fibers as springs which are contracted in the extruder, but expand upon their release.

This fiber-swell is caused by the compression and release of flexible plant fibers acting as leaf springs. Under the pressure of the extruder the fibers are bent. Their force to straighten upon release results in the slurry's expansion. The fiber-swell can then be used to offset the shrinkage caused by the solvent evaporation to reduce or eliminate the overall shrinkage, or possibly to induce an expansion of the extrudate if so desired. The strength of the fiber-swell can be tuned through the fiber dimensions and modulus, as well as through the binder ratio and viscosity. As the shear-thinning causes a fiber alignment along the extrusion bead, both the shrinkage as well as the fiber-swell are reduced along the bead direction, yet act stronger orthogonal to the bead (Fig. 2).

Using the above setup, Kraft-treated softwood and hardwood cellulose fibers were found to not bend significantly and therefore to not cause a fiber-swell, however, untreated softwood fibers at lengths of 10–20 mm did. Figure 2 shows the edge length of a cube of the molded material that expands evenly in all directions after fiber-swell and drying, as well as the radius of an extruded bead, which does not significantly expand or contract along the bead direction but only orthogonal to it. The material expands more with a higher wood content (Fig. 2a), and it expands less with a higher binder viscosity (Fig. 2b). In the combination of treated and untreated fibers (wood content 0.5), to some extent the straightening of the wood fibers appears to overcome the behavior of the cellulose fibers, so that a 1:1 mixture expands more than the mean expansion of the only-cellulose and the only-wood materials.

4.2 Physical Properties

Various material compositions were tested for their compressive and tensile strengths. Those samples had densities ranging from 298 to 549 kg/m3. The tensile yield stress ranged from 0.63 to 13.0 MPa, with Young's moduli in the range of 25.2–428.5 MPa.

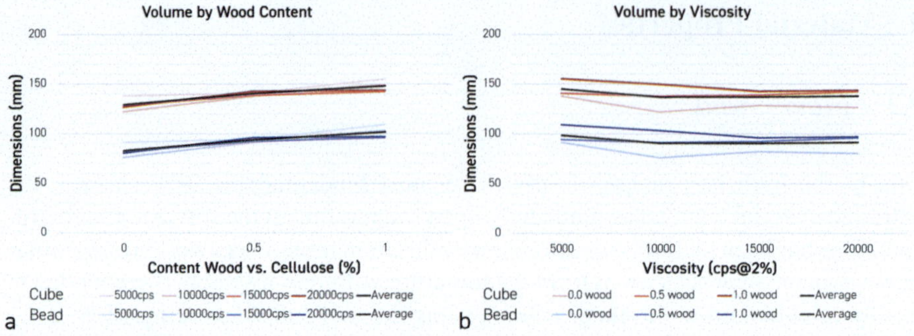

Fig. 2 Effects of the slurry formulation on the volume of the final, cured product. **a** In relation to wood fiber vs. cellulose fiber fraction within the fiber content of the slurry. **b** In relation to the viscosity of the binder. An increase in wood fiber content increases the volume, while an increase in binder viscosity decreases the final volume

The compressive yield stress ranged from 0.33 to 3.22 MPa, with Young's moduli in the range of 24.7–80.7 MPa.

In relation to the fiber expansion, we can observe a decrease in both tensile and compressive yield stress with a decrease in the material density. An increase in wood vs. cellulose fibers therefore leads to increased fiber expansion, a lower density, and lower yield stresses. On the other hand, an increase in binder viscosity leads to less fiber expansion, a higher density, and higher yield stresses (Fig. 3).

5 Case Study

The proposed method has been used for the construction of a large-scale architectural column of 2 m height. The prototype was printed with a slurry of pine wood fibers and cellulose ethers as the binder, thereby consisting of 100% wood products. The wood fibers have dimensions of approximately 10 × 2 × 0.2 mm. The fiber to binder ratio is 60 weight percent, with water used as the solvent. The final dried material has a density of 337.5 kg/m3, with an approximate tensile yield stress of 4.37 mPa and a compressive yield stress of 0.78 mPa.

5.1 Design

The geometry of the column was designed with a differential growth algorithm. Its logic is based on the concept of growth by cell division. A set of nodes, which are connected to their neighbors to form polygonal chains or polygon meshes, has individual nodes iteratively divide and multiply, or be removed from the set, thereby locally increasing or decreasing the length or area of the chain or mesh. Additional

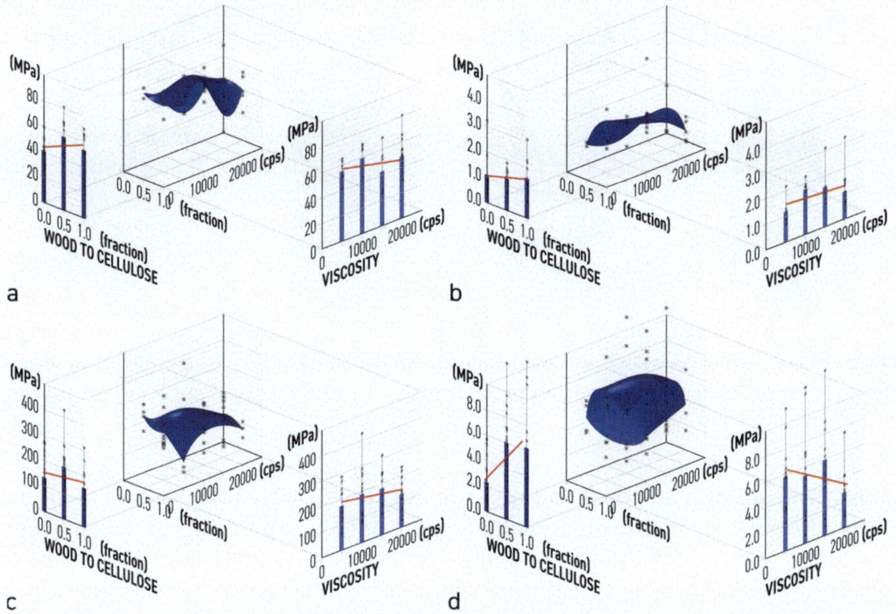

Fig. 3 Effects of the slurry formulation on the physical properties of the final, cured product. **a** tension yield stress. **b** tension Young's modulus. **c** compression yield stress. **d** compression Young's modulus. The regression curves indicate an increase in strength for a higher wood fiber fraction and a decrease for higher binder viscosities

behaviors between the nodes, such as spring forces, attraction or repulsion, can then lead to the formation of complex patterns and geometry, sometimes akin to the formation of grown geometries in nature.

For the design of the prototype, this logic was used to grow trunk-like geometries. Applied to a planar 2-dimensional polygon chain, this chain was allowed to expand and contract over time. The resulting consecutive chains were placed on top of each other to form the individual layers of the 3D print, thereby resulting in the 3-dimensional geometry. An individual node can be constrained to only move within a predefined distance from one iteration to the next. When consecutive chains are placed on top of each other, this can be used to ensure that the 3D geometry does not exceed a given inclination. The maximum inclination thereby equals the maximum allowed distance per iteration, divided by the layer height. The growth logic thereby limited overhangs and ensured the printability of the object (Fig. 4).

5.2 *Toolpath Generation*

Based on the outlines generated by the growth algorithm, which define the outer geometry of the trunk, the tool paths were generated. The printing of discontinuous

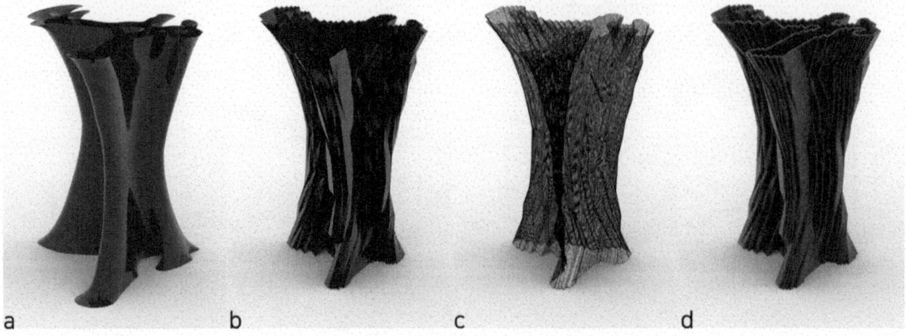

Fig. 4 Design of the prototype. **a** computationally grown target surface. **b** discretization of the target surface to follow defined toolpath angles in plan. **c** toolpath. **d** resulting geometry based on the bead diameter and curvature

toolpaths requires a stopping and restarting of the extruder mid-print. Due to the slurry viscosity and nozzle diameter, a stopping of the extruder did not immediately terminate the flow of the slurry, resulting in undesirable imprecisions. The toolpath was therefore designed as a single, continuously extruded bead.

In its uncured state, the multi-layer extrusion of a single bead thickness wall becomes susceptible to collapse once a certain height is reached. The introduction of waving or kinks in plan can reduce this, especially if the waving increases or decreases with height, so that the printed geometry becomes a double-curved geometry. Furthermore, the extrusion of a cavity wall significantly increases its strength, whereby multiple parallel and interconnected beads with cavities in between form a wider wall.

Therefore, the overall geometry was discretized along defined angles, whereby the originally freeform line segments between the dividing 'cells' were adjusted to always follow those given angles. This resulted in kinks within the longer straight segments of the geometry that increased the extrudate's stability in its wet state (Fig. 4).

In order to generate cavity walls, infill paths were introduced on the interior of the geometry. Instead of beads touching each other in plan, whereby shrinkage during the drying process can result in a delamination of the beads in plan, intersecting beads were found to form better connections. The tool paths were therefore designed in a manner whereby the bead forms a short part of the exterior geometry, it then moves onto the inside and crosses over to the exterior on the opposite side, where it forms another short segment of the exterior, before crossing back over again. In this process, the paths cross several times on the inside and form an integrated structure.

As the bead is a continuous extrusion, and the start and end point of each layer remains in the same location, any cavity wall must have an even number of parallel, intersecting paths, so that the extruder can move back to the start point of a layer along one continuous path. In the case of multi-arm arrangements of walls in plan, the point where two or more arms meet must connect the parallel paths in a way so that the overall toolpath in each layer remains a single continuous curve. Figure 5

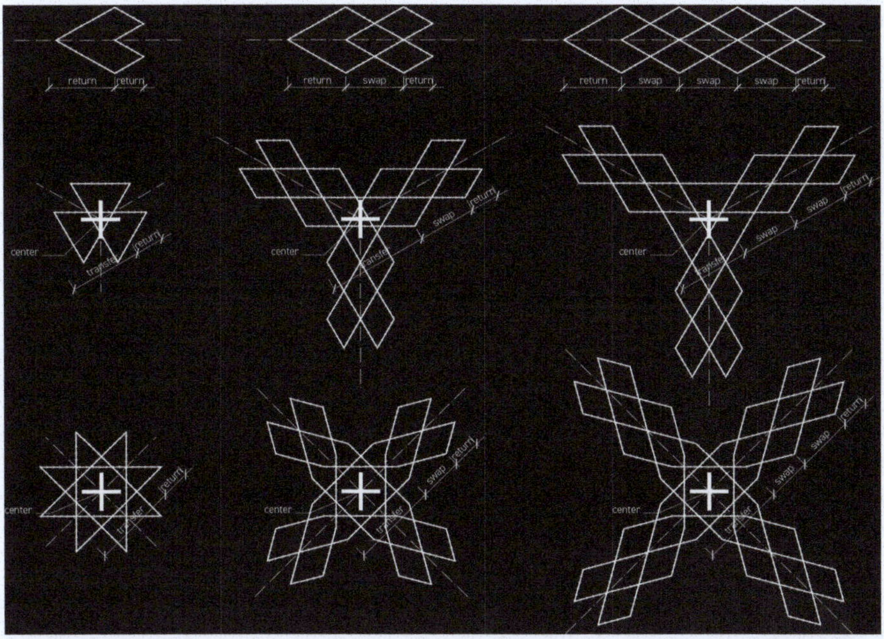

Fig. 5 Diagrammatic development in plan of continuous toolpaths for cavity walls. If the path of a single horizontal layer forms a single, continuous, closed curve, multiple layers can be extruded as a continuous toolpath. Top: Single wall segment of different lengths. Middle: Convergence of three walls while maintaining the continuity of the single curve. Bottom: Convergence of four walls while maintaining the continuity of the single curve

shows examples of those for a single wall, as well as for intersections of three or four walls.

5.3 Discussion

While the construction of the prototype forms a proof-of-concept for the material composition and extrusion method, there are various aspects that offer opportunities for improvements to increase the performance of the printed pieces. Those include aspects of the manufacturing process as well as the final piece.

5.3.1 Hydrolysis and Biodegradation

While the material provides a significant structural performance in the form of wood foams that are light yet strong, hydrolysis is the main drawback of using the material for permanent structural applications. The proposed binders do not chemically cure, but instead form a bond that is reversible on exposure to water. Hydrolysis must therefore be prevented if the pieces are to remain unaffected. While a short time exposure of a printed piece to water has little effect on its structural integrity, a longer exposure will lead to its disintegration. Additionally, biodegradation of the material commences upon a prolonged exposure to water.

The effects of hydrolysis could be prevented by the use of a different binder, especially a chemically curing binder. However, one of the main premises of the project was the aim to create a biodegradable composite. Chemically curing binders mostly result in materials that are not easily biodegradable. Furthermore, hydrolysis often forms the first step in the process of biodegradation, hence it is not completely undesirable for the aims of this project. We are, therefore, reasonably confident in the choice of the category of binder.

A second possibility of preventing hydrolysis and biodegradation could therefore be surface treatments. Most uses of timber in construction involve the use of stains and paints that prevent the wood's exposure to water and subsequent degradation. As the surface of the foam-type material has extensive cavities, the application of many traditional paints is difficult, especially if a continuous surface layer is to be achieved. Also, many paints are often not biodegradable themselves and may cause parts of the final product to not be biodegradable either. However, initial experiments with sodium silicate were promising. On curing, sodium silicate forms silicates. While not biodegradable, silicates nevertheless occur abundantly in nature and are an environmentally benign compound. Sodium silicate could be applied to the wood foams easily. Additionally, a notable increase in the material's physical strength upon treatment with sodium silicate was observed.

5.3.2 Printing Height

The achievable printing heights before collapse or excessive deformations are a concern for the material extrusion of viscous slurries as in the presented method, as has also been observed by other scholar [28, 27]. Depending on the viscosity, the extrudate will form a gel-like structure with limited stability that can collapse under its own weight, while lower layers may be compressed more under the weight of the material above, thereby resulting in wider beads. While the proposed fiber-swell may reduce some of these effects, the process nevertheless must consider those behaviors.

The proposed material composition and method has been used to extrude pieces of up to 500 mm height in a single layer curved geometry, while a height of up to 800 mm could be extruded, but led to irregularities of the geometry that were deemed unusable. An extrusion as cavity walls will lead to an increase in the possible printing height.

The stability of the extrudate can be increased by a reduction in the slurry viscosity, especially by reducing the solvent content. This has the added benefit of a reduction in the material's drying time. However, a lower viscosity slurry will lead to a slower extrusion speed, or the need of a higher torque motor for the extruder. The setup used in the research of this project was constrained in this respect, and for the fabrication

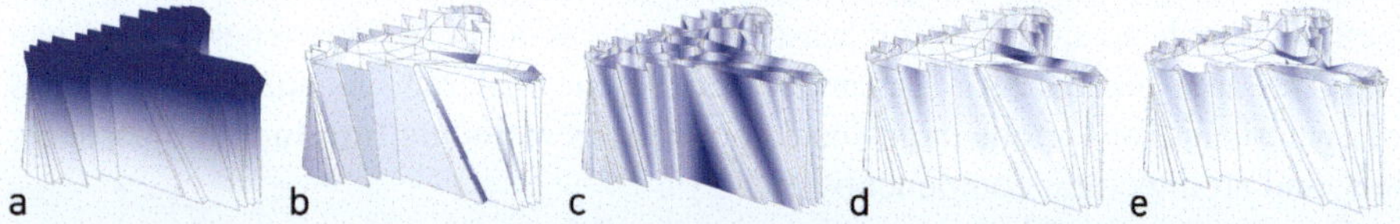

Fig. 6 Estimation of deformations of a printed object under its own weight. **a** distance from the base. **b** wall inclination. **c** distance from the nearest intersection with a crossing wall segment. **d** likelihood of deformation based on the previous factors. **e** Anticipated direction and strength of the deformations

of the larger prototypes, a lower viscosity with higher extrusion speed was preferred. This consequently limited the possible printing height.

5.3.3 Deformations

More specifically, we can estimate the likely deformations that occur based on the geometry to be printed. The three main factors that contribute to the deformation of a point on the geometry are its height above the printing base, the inclination of the object at the specific point, and the distance of the point from its closest intersection with another crossing wall segment. Based on those factors, we can approximate the likeliness of a deformation occurring at a specific location. Furthermore, we can estimate the deformation of the geometry based on the expected displacement at each point, which we assume to be in the direction of the inclination at this point. Those factors and the resulting deformed geometry are shown in Fig. 6.

5.3.4 Surface Finish

While the surface of the untreated extruded pieces has a soft appearance, the material has a high surface hardness, while any protruding wood pieces easily cause scratching. The smoothness of the surface directly relates to the binder viscosity and fiber-swell, as a decreased binder viscosity and increased fiber-swell lead to a coarser foam with more cavities, and to more individual fibers protruding from the material. Beyond the possibility of treating the surface with paints to reduce its roughness, it is also easily possible to sand the dried material.

6 Conclusion

The presented method shows the feasibility of the large-scale 3D printing of wood. Large structures can feasibly be constructed from 100% biodegradable, plant-based materials. The prototype serves as a proof-of-concept, while the testing of samples indicates the significant physical properties of the materials used. The proposed method is able to utilize wood waste material, is low-cost, low-energy, and can produce large objects at a reasonable speed. The material has suitable structural characteristics for various large-scale applications including furniture, molds, stage sets, and architectural components.

Future work should aim to address some of the areas of improvement that the current work has encountered. One main concern is hydrolysis and biodegradation, which occurs upon exposure to water, so chemical or physical means to prevent exposure should be developed without impeding the material's property to biodegrade. Other aspects that warrant further study are an increase of the currently limited printing height, possibly by adjusting the extrusion mechanisms or by tuning the material properties. While the fiber-swell was shown to be able to counteract shrinkage of the material during the drying process, further studies are needed to relate this behavior to the material and extrusion parameters, and to deduct suitable slurry compositions for production. Lastly, post-extrusion treatments such as the proposed spraying with sodium silicate should be explored further to improve the final object's properties.

References

1. Parupelli, S.K., Desai, S.: A comprehensive review of additive manufacturing (3D printing): processes, applications and future potential. Am. J. Appl. Sci. **16**, 244–272 (2019)
2. Tang, M., Shroyer, N.: Cast-in-place freeform concrete with big area additive manufacturing formwork. Int. J. Archit. Eng. Constr. **10**, 1–9 (2021)
3. Leschok, M., Dillenburger, B.: Dissolvable 3DP formwork. In: Bieg, K., Briscoe, D., Odom, C. (eds.) Acadia 2019: Ubiquity and Autonomy: Paper Proceedings of the 39th Annual Conference of the Association for Computer Aided Design in Architecture, Austin, Texas (2019)
4. Love, L., Post, B., Roschli, A., Chesser, P., Hun, D., Brock, S., Brooks, N.: Feasibility of using BAAM for mold inserts for the precast concrete industry (CRADA NFE-17–06874 Final Report). Tech. Rep. (2019)
5. Pegna, J.: Exploratory investigation of solid freeform construction. Autom. Constr. **5**, 427–437 (1997)
6. Khoshnevis, B.: Automated construction by contour crafting—related robotics and information technologies. Autom. Constr. **13**, 5–19 (2004)
7. Cesaretti, G., Dini, E., De Kestelier, X., Colla, V., Pambaguian, L.: Building components for an outpost on the Lunar soil by means of a novel 3D printing technology. Acta Astronaut. **93**, 430–450 (2014)
8. Nematollahi, B., Xia, M., Sanjayan, J.: Current progress of 3D concrete printing technologies. In: ISARC. Proceedings of the International Symposium on Automation and Robotics in Construction, vol. 34 (2017)
9. Weger, D., Gehlen, C., Korte, W., Meyer-Brötz, F., Scheydt, J., Stengel, T.: Building re-thought–3D concrete printing in building practice. Constr. Robot. **5**, 203–210 (2021)
10. Melcher, G., Rauch, A., Yashar, M., Ballard, J.: 3D printing ICONs cosmic pavilion: a case study in cementitious extrusion of unsupported cantilevers. In: Block, P., Boller, G., DeWolf, C., Pauli, J., Kaufmann, W. (eds.) Proceedings of the IASS 2024 Symposium, Zurich, Switzerland. (2024)
11. McGee, W., Ng, T., Yu, K., Li, V.: Extrusion nozzle shaping for improved 3DP of engineered cementitious composites (ECC/SHCC). In: Second RILEM International Conference on Concrete and Digital Fabrication: Digital Concrete 2020, vol 2, pp 916–925. (2020)

12. Zivkovic, S., Lok, L.: Making form work: experiments along the grain of concrete and timber. In: Fabricate 2020: Making Resilient Architecture, pp. 116–123. (2020)
13. Hansemann, G., Schmid, R., Holzinger, C., Tapley, J., Kim, H., Sliskovic, V., Freytag, B., Trummer, A., Peters, S.: Additive Fabrication of Concrete Elements by Robots: Lightweight Concrete Ceiling. In: Fabricate 2020: Making Resilient Architecture, pp. 124–129. (2020)
14. Yu, L.: Arachne—a 3D printed building facade. In: Lamere, J. (ed.) Projects Catalog of the 37th Annual Conference of the Association for Computer Aided Design in Architecture, Boston, pp. 280–285. (2017)
15. Buchanan, C., Gardner, L.: Metal 3D printing in construction: a review of methods, research, applications, opportunities and challenges. Eng. Struct. **180**, 332–348 (2019)
16. Keating, S., Leland, J., Cai, L., Oxman, N.: Toward site-specific and self-sufficient robotic fabrication on architectural scales. Sci. Robot. **2**, eaam8986 (2017)
17. Kariz, M., Sernek, M., Kuzman, M.: Use of wood powder and adhesive as a mixture for 3D printing. Eur. J. Wood Wood Prod. **74**, 123–126 (2016)
18. Kam, D., Layani, M., BarkaiMinerbi, S., Orbaum, D., Abrahami, S., Shoseyov, O., Magdassi, S.: Additive manufacturing of 3D structures composed of wood materials. Adv. Mater. Technol. **4**, 1900158 (2019)
19. Rosenthal, M., Henneberger, C., Gutkes, A., Bues, C.-T.: Liquid deposition modeling: a promising approach for 3D printing of wood. Eur. J. Wood Wood Prod. **76**, 797–799 (2018)
20. Hausmann, M., Siqueira, G., Libanori, R., Kokkinis, D., Neels, A., Zimmermann, T., Studart, A.: Complex-shaped cellulose composites made by wet densification of 3D printed scaffolds. Adv. Funct. Mater. **30**, 1904127 (2020)
21. Deville, S., Saiz, E., Nalla, R., Tomsia, A.: Freezing as a path to build complex composites. Science **311**, 515–518 (2006)
22. Jiang, B., Yao, Y., Liang, Z., Gao, J., Chen, G., Xia, Q., Mi, R., Jiao, M., Wang, X., Hu, L.: Lignin-based direct ink printed structural scaffolds. Small **16**, 1907212 (2020)
23. Thakur, M., Shi, C., Kearney, L., Saadi, M., Meyer, M., Naskar, A., Ajayan, P., Rahman, M.: Three-dimensional printing of wood. Sci. Adv. **10**, eadk3250 (2024)
24. Goidea, A., Floudas, D., Andréen, D.: Pulp Faction: 3d printed material assemblies through microbial biotransformation. In: Fabricate 2020: Making Resilient Architecture, pp. 42–49. (2020)
25. Carcassi, O., Maierdan, Y., Akemah, T., Kawashima, S., Ben-Alon, L.: Maximizing fiber content in 3D printed earth materials: printability, mechanical, thermal and environmental assessments. Constr. Build. Mater. **425**, 135891 (2024)
26. Dritsas, S., Vijay, Y., Halim, S., Teo, R., Sanandiya, N., Fernandez, J.: Cellulosic biocomposites for sustainable manufacturing. In: Fabricate 2020: Making Resilient Architecture, pp. 74–81. (2020)
27. Rossi, G., Chiujdea, R.-S., Hochegger, L., Lharchi, A., Harding, J., Nicholas, P., Tamke, M., Thomsen, M.: Statistically modelling the curing of cellulose-based 3d printed components: methods for material dataset composition, augmentation and encoding. In: Design Modelling Symposium Berlin, pp. 487–500. (2022)
28. Dritsas, S., Ravindran, R., Hoo, J., Fernandez, J.: Shrinkage prediction and correction in material extrusion of cellulose-chitin biopolymers using neural network regression. Virtual Phys. Prototyp. **18**, e2225039 (2023)

Christoph Klemmt graduated from the Architectural Association in London and received his Doctor Technicae from the University of Applied Arts Vienna. He has worked amongst others for Zaha Hadid Architects, where his responsibilities were focused on the company's projects in China. He is Associate Professor at the Department of Design, Art, Architecture and Planning at the University of Cincinnati, where he set up and runs the Architectural Robotics Lab. He co-founded Orproject, an architect's office specializing in advanced geometries with an ecologic agenda. Orproject has been featured widely and has exhibited at the Palais De Tokyo in Paris, the China National Museum in Beijing and the Biennale in Venice.

Gregory Beaucage is Professor of Chemical and Materials Engineering at the University of Cincinnati. He has published extensively in polymers and materials science with 189 publications and 6,900 citations. He has collaborated with Christoph Klemmt on the development of new materials for 3D printing of wood for the past four years. Beaucage holds a Ph.D. from the University of Massachusetts at Amherst in Polymer Science and Engineering and was a post doc at Sandia National Laboratories before joining the faculty at the University of Cincinnati. He was a Fulbright Global Scholar at Sheffield University in the UK and at several Universities in Ethiopia, South Africa, and Lesotho. He has been Visiting Professor at University of the Witwatersrand, Johannesburg, University of Sheffield, University of Cape Town, and ETH in Zürich.

Chapter 11
3D Printing with Living Matter

Nancy Diniz and Frank Melendez

Abstract This chapter explores the theoretical implications of co-authorship with living systems in the context of bio-digital fabrication. It presents a body of work focused on 3D printing and robotic deposition techniques that integrate living systems such as fungi, lichens, algae, and bacteria. These microorganisms are embedded into bioreceptive, hydrogel-based substrates, forming hybrid material systems that grow, adapt, and eventually decompose—challenging traditional design notions of form, function, permanence, and authorship. Grounded in the concept of bioreceptivity, the research investigates how computational design, microbiology, and digital fabrication can converge to develop living materials that function as active ecological interfaces. These materials are not static but evolve through time and environmental interaction, becoming carbon-positive structures that heal, grow, and regenerate. The methodologies combine generative modeling, environmental simulation tools, and machine learning with lab-based protocols for cultivating and manipulating living cells. These living materials operate within circular economies, utilizing local, renewable resources and upcycled waste, while reducing energy and carbon footprints. By treating biological systems as collaborators rather than passive materials, this research proposes new frameworks for regenerative design practice. It emphasizes low-energy, toxin-free, biodegradable lifecycles, offering alternatives to extractive, high-waste industrial processes. Ultimately, the work repositions fabrication as an act of cultivation and care, where the designer engages in a dynamic, symbiotic relationship with material life. This approach contributes to the growing discourse on biodesign and living systems thinking, promoting a shift toward ecologically responsive and regenerative design strategies embedded in both scientific rigor and creative experimentation.

N. Diniz (✉) · F. Melendez
bioMATTERS, New York, NY, USA
e-mail: nancy.diniz@gmail.com

N. Diniz
Central Saint Martins, University of the Arts London, London, United Kingdom

F. Melendez
Spitzer School of Architecture, City College of New York, CUNY, New York, NY, USA

© The Author(s), under exclusive license to Springer Nature Singapore Pte Ltd. 2026

C. Klemmt and G. Beaucage (eds.), *Biodegradable 3D-Printing*, The Materials Research Society Series, https://doi.org/10.1007/978-981-95-4025-9_11

Keywords Biomaterials · Biodesign · Biofabrication · 3D printing mycelium · Robotic fabrication · Living materials · Regenerative design · Bio-digital craft

1 Introduction

This chapter presents a body of research by our studio bioMATTERS, focused on bio-digital fabrication, with particular emphasis on 3D printing and robotic deposition techniques that incorporate living microorganisms such as fungi, bacteria, and microalgae. These organisms are embedded within hydrogel-based substrates and integrated into additive manufacturing processes to create biodegradable, bioreceptive materials—materials that actively support biological colonization and exhibit environmental responsiveness. The chapter will first introduce a series of design projects that integrate life organisms and then discuss the historical and theoretical framework of this approach. The approach is rooted in the concept of bioreceptivity, first introduced by the Belgian environmental scientist, Olivier Guillitte, which defines a material's inherent capacity to be colonized by living organisms [1]. In this context, bioreceptive materials are intentionally designed to host and sustain microbial life, allowing organisms such as algae, fungi, or bacteria to interact with and transform their environment. Building on this foundation, the work described in this chapter explores the convergence of computational design, digital fabrication, and biological sciences to develop novel living material systems. Through interdisciplinary workflows, the research brings together laboratory protocols for cultivating and manipulating living cells with advanced computational design strategies. It incorporates environmental simulation tools, generative modeling techniques, and robotic 3D printing technologies, alongside expertise in microbiology and materials science. This convergence enables the development of bio-integrated fabrication methods that not only foster biological growth but also exhibit adaptive, responsive behaviors in relation to their surroundings. By framing bio-digital fabrication as a platform for collaborative material agency, this research repositions the role of the designer in an active dialogue with living processes. Rather than producing static objects, design becomes a practice of orchestrating dynamic systems that unfold over time, guided by both human intent and biological rhythms in which we explore 3D printing with living cells. (Fig. 1). This chapter examines the theoretical and historical implications of utilizing living organisms in the field of design, and further, contributes to ongoing discussions in biodesign by addressing the aesthetic, ethical, and ecological dimensions of working with living matter. It examines how design—conceptually, functionally, and temporally—can evolve toward a symbiotic future, where living processes are not merely accommodated but actively embraced as central to the design and material culture of the built environment.

Fig. 1 bioMATTERS, mycelium vessels, 2022–2023. computationally designed, 3D printed mycelium-based composites

2 Bioreceptive Materials

In order to situate practices within topic living materials research, an overview of notable work exploring the interdisciplinary field of 3D printed bioreceptive materials employing various microorganisms is presented. Research into living materials encompasses the study of bioreceptivity. Bioreceptive materials can be designed to host and nourish various living organisms, such as microbes, fungi, or algae, allowing them to grow and interact with their environment. The integration of living cells into 3D printing for design-oriented applications builds upon foundational research in biomedical engineering. Researchers at the Wake Forest Institute for Regenerative Medicine, Sean V. Murphy and Anthony Atala, offer a comprehensive review of 3D bioprinting technologies used to fabricate tissue engineering scaffolds, highlighting hydrogels and composite biomaterials for replicating complex biological tissues [2]. While developed for medical applications, their work lays important groundwork for the cross-pollination of biological and design disciplines.

Expanding into art, architecture and design, numerous researchers have developed interdisciplinary workflows involving microbial cultures and digital fabrication tools to create responsive, bioreceptive systems. Rachel Armstrong, Professor of Regenerative Architecture at KU Leuven, Belgium, has pioneered research into living architectural systems, particularly in the use of microbial processes and biocomposites for self-sustaining environments. Her work on photosynthetic textiles demonstrates the integration of microalgae into flexible substrates, producing active materials capable of CO_2 sequestration and dynamic environmental responsiveness—proposing architecture as a metabolically active system [3]. These biocomposites aim to contribute to the greening of the built environment, offering ecological benefits such as air purification. At the Bartlett School of Architecture, Professors Marcos Cruz and Richard Beckett explored the concept of bioreceptive concrete, cast into CNC milled molds, to support microbial colonization, including mosses, algae, and biofilms [4]. By engineering specific surface topographies and porosities, they propose a regenerative design methodology, transforming building facades into ecological substrates that enhance biodiversity and air quality in dense urban environments. This work positions bioreceptive materials as a transformative strategy for regenerative building systems. Work by the designers and educators, Ronald Rael and Virginia San Fratello—particularly in their book *Printing Architecture* [5]—has inspired new material experimentation and digital craft, grounded in ecological responsibility. They reframe 3D printing as a means of reinventing architecture through a range of materials, including sustainable materials, such as clay, sand, bioplastic, and clay, to name a few. Expanding upon this work includes 3D printed bioreceptive ceramics [6, 7].

The integration of 3D printing technology with mycelium-based materials has garnered significant attention in design and architecture, leading to innovative applications that leverage the unique properties of fungal mycelium. The seminal project, "Mycelium Chair" by Dutch designer, Eric Klarenbeek, represents the first concept for 3D printed fungi [8]. The concept chair was a 3D printed bioplastic monocoque shell that was then filled with a substrate colonized with living mycelium which grew, adding structural strength to the chair. Other 3D printed mycelium-based work has been developed by researchers Elise Elsacker, Eveline Peeters, and Lars De Laet [9, 10] and bioMATTERS [11]. These works explore the potential for 3D printing and robotic extrusion-based additive manufacturing using mycelium-based composites to create sustainable, biodegradable, and structurally stable bio fabricated structures.

A more recent advance comes from the Hub for Biotechnology in the Built Environment, where Martyn Dade-Robertson and Meng Zhang developed *Living Manufacture*, a microbial 3D printing system that utilizes bacterial cellulose grown in situ from genetically engineered microbes [12]. Their work involves the creation of a novel bioreactor integrated into a new 3D printer platform to fabricate biologically grown, functionally graded materials. This innovation marks a shift toward biological manufacturing systems, with implications for regenerative design, high-performance composites, and consumer product innovation. Together, these projects represent an emergent paradigm in which digital fabrication, synthetic biology, and microbial agency converge—redefining the aesthetics, performance, and sustainability of man-made environments.

These various workflows rely on interdisciplinary collaborations, integrating expertise in design, science, and technologies. They merge computational design, environmental analysis tools, digital fabrication, and design engineering, with biological sciences that integrate lab protocols and biointegration strategies. This convergence enables the development of materials and structures that foster new possibilities for biologically grown and environmentally responsive designed objects and systems.

3 Prototypes

3.1 3D Printing with Living Organisms

3D printing with biological systems requires examining the performance of the materials under different incubation conditions to support the metabolic processes of the microorganisms involved. This can vary widely, depending on various species of microalgae, fungi, lichens, bacteria, or plant seeds. This research investigates how biocomposites can be fabricated using digital technologies, emphasizing the importance of combining biomaterials with advanced manufacturing techniques such as 3D printing to create more environmentally responsive and sustainable building materials.

Prior to incorporating living organisms, the research and work at our studio, bioMATTERS, focused on developing 3D printing techniques using waste-based substrates. Both previous and current work emphasizes upcycling—transforming industrial and domestic waste into high-value materials. This diverges from recycling by adding value through design and reducing landfill waste, especially in the Global South. Notable examples of companies that focus on upcycled biomaterials include pineapple-based leather (Piñatex), coffee-ground sneakers (Rens), and mango leather (Fruitleather Rotterdam) [13]. Recent work continues to focus on upcycled waste materials, to develop mycelium-based composites, centered on bio-digital fabrication techniques that incorporate 3D printing and robotic deposition. This work results in biodegradable structures for use in furniture, interiors, and architectural systems.

Similar to the ancient process of forming ceramic bowls and vessels using the technique of clay coiling, 3D printing offers a similar process of building up layers of material to form objects. Our series of mycelium bowls and vessels are inspired by these ancient techniques and traditional methods of forming pottery to create domestic and utilitarian objects. Unlike traditional hand-made craft, these designs use parametric models and generative algorithms to determine their formal qualities. (Fig. 2). Additionally, the material agency of the mycelium is expressed through emergent surface textures, patterns, and color. (Fig. 3).

Fig. 2 bioMATTERS, mycelium bowl, 2024. Computationally designed, 3D printed mycelium-based composites

3.2 3D Printing with Mycelium

The *MYCO-CLAY* vessels combine clay and mycelium-based composites, both abundant materials that can be locally sourced. Clay, a natural material used by humans dating back multiple millenia, is one of the most ancient materials used for making objects, tools, art, and architecture. As a natural material, clay offers a more sustainable approach to design, as opposed to synthetic and environmentally detrimental materials that are used in the design and fabrication of products and buildings. Additionally, the availability and abundance of clay worldwide allows for the design and fabrication of ceramics that can be made with locally sourced clays, and the potential to address cultural traditions and histories. These vessels use differential growth algorithms to create structurally stable forms that result in various sinuous geometries and continuous surface undulations. (Fig. 4). Here, the combination of mycelium growth gradually transitions from the natural, off-white color of mycelium, to the natural color of the terra-cotta earthenware clay, binding layers together and serving as an aesthetic enhancing bio-glaze. (Fig. 5).

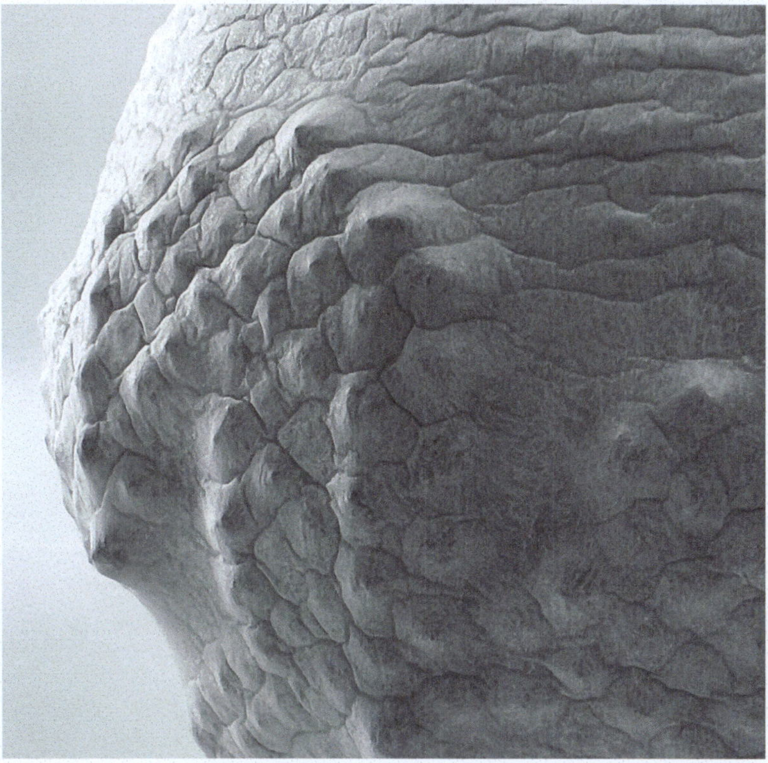

Fig. 3 bioMATTERS, mycelium bowl, 2024. Detail of emergent mycelium surface texture, pattern, and color

The *MYCO-ALGA* architectural panels use an L-system branching logic to create unique morphologies inspired by mycelial growth. Unlike casting techniques, which require molds to create identical parts, 3D printing allows for the fabrication of bespoke panels, each reflecting a unique pattern. This results in continuous variation when assembled across larger surfaces. In addition to the mycelium material, bio-pigments extracted from a harvested biomass of algae are made into gels that are printed onto each panel, adding natural color as an aesthetic enhancement. The finished architectural tiles are designed to enhance interior spaces, displaying the natural off-white color of mycelium and light green algae, while offering a soft, velvet-like texture and tactile quality. (Fig. 6).

These biohybrid objects—partly grown, partly fabricated—merge algorithmic logic with biological growth. Inspired by natural growth patterns and forms, we use algorithms and computational simulations to generate formal complexity and performative attributes. Digital models are then translated into 3D printable geometries, where parameters such as x, y, z coordinates, print speed, and extrusion pressure are integrated into generative workflows. Post 3D printing, the designed objects are placed into sterile environments, where mycelium grows, binds, and transforms the

Fig. 4 bioMATTERS, *MYCO-CLAY* vessel, 2023. 3D printed mycelium-based composites and earthenware clay

material, yielding unique patterns and textures that manifest a contemporary bio-aesthetic. Mycelium responds to nutritional and environmental factors, such as light, temperature, and humidity, influencing both its growth and material properties. Its biodegradability, light weight, and carbon neutrality make it an ideal candidate for sustainable material assemblies. This approach aims to catalyze a cultural shift toward design that centers ecological intelligence and bio-digital craft.

3.3 3D Printing with Lichens

In the project *Symbiont Substrate*, we tested the potential for clay 3D printing at a larger scale, as a bioreceptive substrate for lichen colonization. Lichens are a symbiotic organism consisting of fungi and algae or cyanobacteria. In their mutualistic relationship, the fungus provide structure and nutrients, while the algae or cyanobacteria provide nutrients through photosynthesis. Lichens grow on various surfaces, or substrates, that are natural, such as trees and rocks, as well as human-made, such as

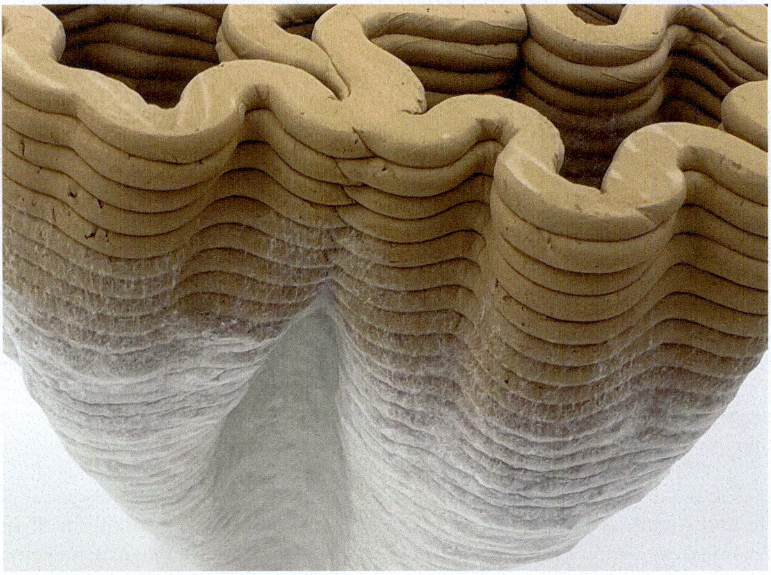

Fig. 5 bioMATTERS, *MYCO-CLAY* vessel, 2023. Detail of mycelium hyphae filaments gradation and growth over earthenware clay

Fig. 6 bioMATTERS, *MYCO-ALGA* architectural panels, 2022–2023. Computationally designed, 3D printed mycelium-based composites with algae based pigments

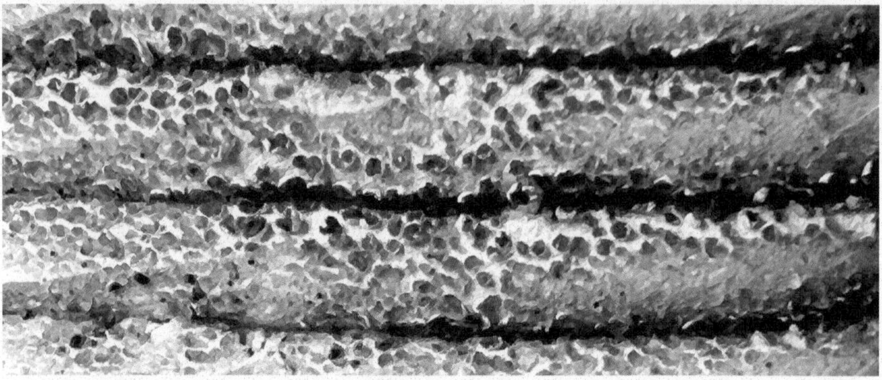

Fig. 7 bioMATTERS, *Symbiont Substrate*, 2023. Detail of 3D printed clay/ceramic substrate with lichen colonization

concrete and metal. Additionally, the research explores the colonization of lichens on 3D printed ceramics elements using a robotic arm, as a means of merging bio-digital fabrication workflows. Initially, a series of clay 3D printed modules were developed, and bisque fired at a low temperature of cone 04 (1940 °F/1060 °C), to create ceramic modules for testing lichen colonization and growth. These growth studies were conducted with different variables, including lichen species, growth media, light, and moisture levels. After several weeks, lichen growth was evident on the 3D printed ceramic substrate. (Fig. 7).

The project consists of a series of 3D printed ceramic components that are assembled and stacked vertically to form an architectural column. (Fig. 8). The form and surface textures merge computational precision with the unruly and dynamic characteristics of clay. (Fig. 9). An industrial robot was used to develop a biofabrication workflow to colonize the ceramic column with lichen spores. A customized robotic end-effector held a spray gun with lichen ink to precisely position the lichens and pattern on the surface and create continuity between the modules. The various pockets, crevices, blobs, and folds are intended to create a range of various micro-climatic conditions for lichens to adapt and thrive. The resulting design explores the aesthetics of bio-digital craft, blurring physical material with digital form and a controlled looseness that results in a post-digital material expression [14]. *Symbiont Substrate* is a prototype that reconsiders architectural orders through an ecological lens that promotes biodiversity in urban environments. Additionally, the lichens, which are sensitive to air pollution, are intended to serve as indicators of air quality, as urban bio-sensors.

Our design approach leverages multiple types of microorganisms—including fungi, algae, lichens, and bacteria—all of which contribute to low-impact, biodegradable material systems. In tandem with upcycled waste materials and natural materials, such as clay, these systems transform surfaces into active climate solutions by developing carbon-positive, regenerative, and naturally decomposing design strategies. The methodology reflects an interdisciplinary workflow that integrates microbiology,

Fig. 8 bioMATTERS, *Symbiont Substrate*, 2023. Assembly and detail of 3D printed architectural column assembly

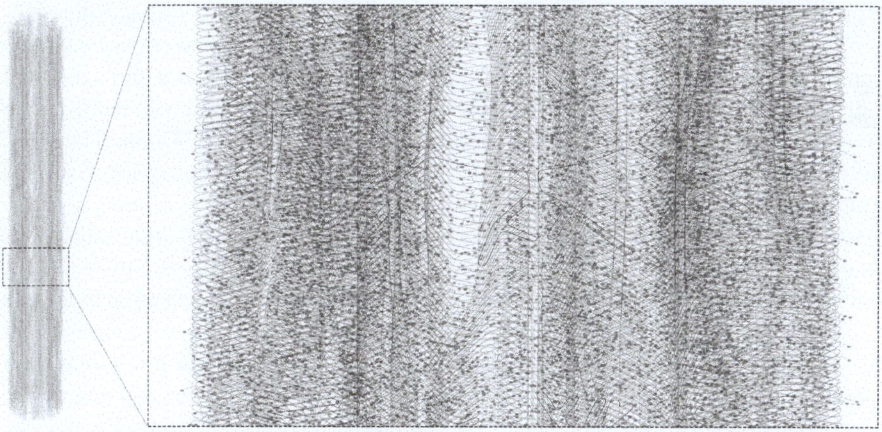

Fig. 9 bioMATTERS, *Symbiont Substrate*, 2023. Drawing of the toolpaths to generate the G-code for 3D printing

computational design, and digital fabrication to guide and engineer living systems. Our projects attempt to blur the boundaries between the natural and artificial, with some employing machine learning and AI to interpret biological data and generate bio-morphic forms. In mycelium-based design, the shift is not merely from artifact to process, but from static form to temporal, living systems. Using 3D printers and robotic arms, we scale up biofabrication while balancing structural integrity and biological viability. We investigate the theoretical implications of integrating 3D printing with living cells into the design and fabrication of human-made objects. These methods open new possibilities for architectural speculation, marking a shift from traditional systems thinking to living systems thinking.

4 Discussion

4.1 The Shift Toward Nonhuman Design

The term *nonhuman* was introduced by Richard Grusin in the early 2010s, particularly in the context of the 2012 "Nonhuman Turn" conference at the University of Wisconsin-Milwaukee [15]. The theory of the *Nonhuman* marks a significant shift within the humanities and social sciences, wherein the focus moves away from anthropocentrism toward an acknowledgment of nonhuman agency, materiality, and interrelations. This approach spans fields such as philosophy, media studies, and design, and draws from several influential intellectual frameworks. A foundational influence is Bruno Latour's *Actor-Network Theory*, which conceptualizes agency as distributed across both human and nonhuman entities [16]. This is extended through his notion of *compositionism*, which calls for building and sustaining networks through a deeper intimacy and care for nonhuman actors [17]. Related developments are found in *New Materialism* theories, particularly in the works of scholars Jane Bennett and Karen Barad, whose theories emphasize the vitality and agency of matter, and the entanglement of material and meaning [18]. These perspectives are pivotal in posthumanist and feminist science studies. The scholar Donna Haraway, another key figure, has been instrumental in challenging human exceptionalism and exploring multispecies entanglements. Her seminal work *A Cyborg Manifesto* introduced the cyborg as a hybrid entity—part machine, part organism—that transcends binary divisions such as nature/culture and human/machine [19]. Haraway's work continues to inspire designers to rethink materiality, ethics, and co-creation with both living and non-living systems, significantly shaping discourse around future ecologies and multispecies design. Within design and architecture, the nonhuman turn calls for reimagining the built environment as co-produced by human and nonhuman actors—ranging from microbes and materials to ecosystems and digital networks. These practices signal a shift from the Anthropocene toward the *Ecocene* [20], where design engages natural forces and co-creates futures grounded in mutual flourishing. This framework supports the emergence of techno-biological design approaches that prioritize the creation of multispecies environments, aiming to be viable not only for humans but for all forms of life. It opens speculative avenues for future ecologies, enabling strategies that acknowledge our entanglement within broader biological and technological systems.

4.2 The Emergence of Bio-Digital Craft

At the core of our methodology lies the integration of computational design tools with biologically derived materials, which allows for high levels of customization at both the material and form scale. Unlike conventional fabrication, 3D printing and robotic deposition can respond to variable material properties, such as viscosity, shrinkage, or growth behavior—factors inherent to biomaterials like mycelium, bacterial cellulose, or alginate-based gels. One of the major limitations of working with living or degradable materials is the challenge of scaling up without losing precision or performance. Robotic biofabrication overcomes this by enabling continuous deposition of bio-based pastes, gels, or composite mixtures onto scaffolds or directly into formwork, facilitating the construction of larger structural or surface elements without manual intervention.

Computer-Aided Design (CAD), encompassing 3D modeling and digital fabrication through CNC machines, has a rich lineage. Its origins can be traced to mid-20th-century cybernetics, notably in the work of Norbert Wiener, Gray Walter, and Gordon Pask. Walter's *Machina Speculatrix*, a series of autonomous robots that reacted to stimuli, exemplified early efforts to simulate adaptive behavior [21]. Initially, a postwar science of the adaptive brain, emerging from the field of psychology, cybernetics expanded to influence many other fields, including robotics, politics, engineering, architecture, music, and more. This includes the architectural visions of British architect Cedric Price, and British theater director Joan Littlewood's, *Fun Palace*—a structure conceived to adapt and respond to user input. Subsequent projects, such as *Walking City* by Archigram, and *SEEK* by Nicholas Negroponte (MIT Media Lab), continued to push the boundaries and explore the possibilities of automated computer controlled architectures that could respond, communicate, and provide feedback [22]. By the 1990s and early 2000s, architects increasingly adopted digital tools, including parametric design, algorithmic generation, and environmental simulation. These technologies led to architectural theories that were based on biological principles through the design of forms that were organic, curvilinear and evolutionary. Architects and theorists such as Greg Lynn explored biological concepts through animation software to inspire forms generated by forces [23], while Michael Weinstock investigated morphogenesis and self-organization through evolutionary algorithms [24].

Digital fabrication tools evolved in tandem, allowing designers to physically output complex biologically inspired 3D models using high precision 3D printing, CNC milling, and robotic fabrication technologies. Early notions of digital fabrication as a means of creating high precision replicas of digital models evolved to incorporate other material qualities and expressions. This trajectory reflected a re-engagement with traditional notions of craft, where risk and uncertainty in working with materials are intrinsic to the process [25], leading toward post-digital notions of craft, where material agency blurs the boundaries between digital-physical and precision-imprecision [14]. Within a post-digital fabrication approach, machines, 3d printers, and robots become an extension of the craftsperson. Tooling, toolpaths, and G-code settings of pressure, speed, coordinates, and extrusion rates become variables for design calibration with material feedback.

Bio-digital fabrication processes expand upon these concepts by working with actual biological systems, materials, and processes, in an effort to embrace multi-species collaborations that promote sustainable and ecological design solutions. The architect and designer, Neri Oxman, coined the term "material ecology" in an aim to establish deeper relationships between designed objects and their environments by merging multifunctional materials, fabrication, and computation [26]. Our research and work builds on these concepts, working with various living organisms, upcycled materials, algorithmic processes, and 3D printing—robotic deposition, to co-design and co-fabricate objects that promote ecological practices. This is in response to negative environmental impacts brought on by Anthropocentric practices, which have led to natural resource depletion and the climate crisis. Combined with computational design and additive manufacturing tools, these methods offer novel bio-digital workflows that reflect the complex interconnected relationships between machines, materials, environments, and biological systems. They combine biological intelligence and organic materials to advance traditional notions of craft.

5 Conclusion

This chapter presents the use of 3D printing and robotic biofabrication processes, to serve not only as production tools, but as co-evolutionary systems that align material agency with design intent. These processes enable the mass-customization of ecologically embedded designs, scaling the potential of bio-based materials from lab to design scale—without relying on extractive, high-waste manufacturing paradigms. This synergetic approach aims to propose not only new material aesthetics but new material economies for product design, art and architecture.

The topic of bio-digital design opens new paradigms for rethinking fabrication, materiality, and the act of designing with living systems. Working with living and biodegradable materials reorients conventional design values—conceptual, functional, temporal, and aesthetic—toward the promotion of a decarbonized, regenerative, and symbiotic future. In the future, biological rhythms and metabolic processes are not only embraced but become integral to the creative act. Design shifts from producing static artifacts to orchestrating distributed, time-based processes that unfold and evolve in dialogue with their environments. Additive manufacturing and computational design technologies now enable new modes of bio-digital craft—methods that combine precision and automation with biological responsiveness.

Our research contributes to this growing field through the development of 3D printed bioreceptive substrates that support the colonization of various microorganisms, including fungi, lichens, algae, and bacteria. Each of these organisms offers unique capabilities—biodegradability, low-energy input, toxin-free lifecycles—that contribute to environmentally sensitive fabrication. Through these systems, surfaces become active ecological interfaces, transforming into carbon-positive structures that grow, heal, and ultimately decompose naturally. The methodologies developed reflect a cross-disciplinary approach that merges microbiology, computational design, and digital fabrication to guide and shape the growth of hybrid materialities—partially designed, partially living. Some of these projects employ generative algorithms and computer simulations to model and influence biological behavior, while others use machine learning and artificial intelligence to interpret biological data and generate responsive, bio-informed morphologies. This body of work contributes to a theoretical discourse around the future of design practice in co-authorship with living systems. By embedding intelligence, adaptability, and ecological sensitivity into material systems, biodesign challenges us to reimagine fabrication as a process of cultivation, co-existence, and care. It examines how the integration of living and biodegradable materials prompts a rethinking of form, function, temporality, and material lifecycle—challenging conventional notions of authorship, permanence, and control in design. By treating living systems as active collaborators rather than passive materials, this research proposes new frameworks for design practice that are more attuned to the complexities and interdependencies of ecological systems.

This approach is rooted in scientific research, computational design and cultural heritage, bringing together fieldwork, environmental data, and traditional craftsmanship. Integrating living organisms in design enables a paradigm shift—from passive, extractive materials to *active, regenerative* systems that participate in ecological cycles. Living systems have the intrinsic ability to grow using renewable resources, often from waste. These organisms don't require high heat, toxic chemicals, or intensive mining, and they can often be grown locally using closed-loop systems, which drastically reduces embodied energy and carbon footprint. In terms of durability, products are designed for decomposition or long-term use. Unlike conventional materials that aim for permanence and often become waste, living materials can be engineered with time in mind. This intentional design of material lifecycles fosters a circular economy, where materials return safely to the environment or regenerate through biological processes. Living organisms also bring functional qualities that conventional materials cannot replicate. For example, algae can photosynthesize, making it possible to develop surfaces that purify air or generate oxygen and biomass. Bacteria can be designed to respond to environmental cues—such as releasing scent, pigment, or antimicrobial substances under specific conditions. Mycelium has natural insulating, fire-resistant, and acoustic-dampening properties, making it a compelling alternative for construction. These materials don't just *exist* in space—they can interact, adapt, and evolve, enabling new kinds of responsive and symbiotic architectures.

References

1. Guillitte, O.: Bioreceptivity: a new concept for building ecological studies. Sci. Total. Environ. **167**(1–3), 215–220 (1995)
2. Murphy, S.V., Atala, A.: Recent advances in 3D printing of biomaterials. J. Biol. Eng. **9**(1), 1 (2015)
3. Armstrong, R., Stefanova, A., Inna, P., Caldwell, G.S.: Bridgens B: photosynthetic textile biocomposites: using laboratory testing and digital fabrication to develop flexible living building materials. Sustain. Environ. Manag. Constr. **5**(1), 1–10 (2021)
4. Cruz, M., Beckett, R.: Bioreceptive design: a novel approach to biodigital materiality. Archit. Res. Q. **20**(1), 51–64 (2016)
5. Rael, R., San Fratello, V.: Printing Architecture: Innovative Recipes for 3D Printing. Chronicle Books (2018)
6. Crawford, A., et al: Clay 3D printing as a bio-design research tool: development of photosynthetic living building components. Archit. Sci. Rev. **65**(3) (2022)
7. Diniz, N., Melendez, F: Symbiont substrate: a bio-digital fabrication method for large scale 3D printing of bioreceptive clay modules and automated colonization of lichen communities. In: Proceedings of the Association for Computer-Aided Design in Architecture (ACADIA) (2023)
8. Fairs, M.: Mycelium chair by Eric Klarenbeek is 3D-printed with living fungus. Dezeen. https://www.dezeen.com/2013/10/20/mycelium-chair-by-eric-klarenbeek-is-3d-printed-with-living-fungus/ (2013). Accessed 12 Apr 2025
9. Elsacker, E., et al: A comprehensive framework for the production of mycelium-based lignocellulosic composites. Sci. Total Environ. **725** (2020)
10. Elsacker, E., Peeters, E., De Laet, L: Large-scale robotic extrusion-based additive manufacturing with living mycelium materials. Sustain. Futures **4**, 100085 (2022)
11. Diniz, N., Melendez, F: Inoculated matter: 3D printing with mycelium and upcycled waste. In: Proceedings of the Association for Computer-Aided Design in Architecture (ACADIA) (2021)
12. Dade-Robertson, M., Arnardottir, T.H., Lee, S., Loh, J., Gilmour, K., Zhang, M.: Engineered living fabrication: combining hardware, wetware and software for the non-entropic guided growth of microbial cellulose. In: Crawford, A., Diniz, N., Beckett, R., Vanucchi, J., Swackhamer, M. (eds.) Habits of the Anthropocene: Scarcity and Abundance in a Post-Material Economy. Proceedings of the 43rd Annual Conference of the Association for Computer Aided Design in Architecture (ACADIA 2023), Denver, USA, 21–28 October 2023, pp. 194–204 Acadia Publishing Company, Mount Pleasant (2023)
13. Tewari, S., Reshamwala, S.M.S., Bhatt, L., Kale, R.D: Vegan leather: a sustainable reality or a marketing gimmick? Environ. Sci. Pollut. Res. **31**(3), 3361–3375 (2024)
14. Fure, A: Digital materiallurgy: on the productive force of deep codes and vital matter. In: Proceedings of the Association for Computer-Aided Design in Architecture (ACADIA) (2011)
15. Grusin, R.: The Nonhuman Turn. University of Minnesota Press, Minneapolis (2015)
16. Latour, B.: Reassembling the Social: An Introduction to Actor-Network-Theory. Oxford University Press, Oxford (2005)
17. Latour, B.: Love Your Monsters: Why We Must Care for Our Technologies as We Do Our Children. Breakthrough J. **2**. https://thebreakthrough.org/articles/love-your-monsters (2012). Accessed 12 Apr 2025
18. Bennett, J.: Vibrant Matter: A Political Ecology of Things. Duke University Press, Durham, NC (2010)
19. Haraway, D.J: A cyborg manifesto: science, technology, and socialist-feminism in the late twentieth century. Social. Rev. **80**, 65–108 (1985)
20. Armstrong, R: Transitioning towards the Ecoscene. In: Imhof, B., Gruber, P. (eds.) Built to Grow: Blending Architecture and Biology. Birkhäuser, Berlin, Boston (2015)
21. Pickering, A.: The Cybernetic Brain: Sketches of Another Future. University of Chicago Press, Chicago and London (2010)
22. Sadler, S.: Archigram: Architecture Without Architecture. Mit Press (2005)

23. Lynn, G.: Animate Form. Princeton Architectural Press, New York (1999)
24. Weinstock, M.: The Architecture of Emergence: The Evolution of Form in Nature and Civilization. John Wiley & Sons, West Sussex, UK (2010)
25. Pye, D.: The Nature and Art of Workmanship. Cambridge University Press, Cambridge (1968)
26. Oxman, N., et al: Material ecology. Comput. Aided Des. **60**, 1–2 (2015)

Nancy Diniz is co-founder of bioMATTERS, LLC and Course Leader of MA Biodesign at Central Saint Martins, UAL. She has exhibited and curated symposia on living materials, bio-digital craft, and interactive design. Her work has been shown at Grand Designs Live, Material Matters, Lisbon Architecture Triennale, MAAT Museum, and Istanbul Design Biennial. She has received awards, fellowships, and residencies from Heritage Crafts QEST, Haystack, Museum of Architecture and Design (MAO) / Center Rog Slovenia, Somerset House, NEW INC, MacDowell, and EYEBEAM. Nancy co-edited Data, Matter, Design (Routledge, 2020) and edited Bio-Calibrated (Cambridge University Press, 2023). She holds degrees in architecture and computational design from Lusiada University and The Bartlett School of Architecture UCL.

Frank Melendez is co-founder of bioMATTERS, LLC and the Department Chair and an Associate Professor at the Spitzer School of Architecture, CCNY, CUNY. He is the author of Drawing from the Model (Wiley, 2019) and co-editor of Data, Matter, Design (Routledge, 2020). His work has been exhibited at London Design Festival, Usagi Gallery, Lisbon Architecture Triennale, and Today Museum. He has curated symposia and exhibitions, including rE-Ordering Architecture (NY) and DATA & MATTER (Venice). Frank has received grants and residencies from Haystack, Museum of Architecture and Design (MAO) / Center Rog Slovenia, Somerset House, NYSCA/Van Alen, NAHR, MacDowell, and NEW INC. He holds degrees in architecture from the University of Arizona and Yale University.

Part IV
Conclusions

Chapter 12
Conclusions

Christoph Klemmt and Gregory Beaucage

Abstract This conclusion provides a recapitulation of the chapters of this book on biodegradable 3D printing, based on the three sections of Introduction, Materials and Methods, and Applications. It then highlights several aspects that have been pointed out by different chapters, including common findings regarding the materials used, suggested applications, biodegradability, and challenges to the proposed methods. Several chapters raised questions regarding issues such as the materials' biodegradation, design of geometry and toolpaths for a continuous deposition, concerns regarding collapse of uncured extrudates, shrinkage and deformations, viability of bioprinting, and applicability and commercialization. This conclusion then points out some of the important areas for future development that are deemed of relevance to advance the field of 3D printing with biodegradable materials.

Keywords 3D printing · Biodegradable · Materials · Additive manufacturing · Sustainability

1 Introduction

3D printing has seen large growth and development over the past decades, providing ever better methods, precision, scale, and a range of novel materials and applications. However, while the world is moving toward increased sustainability, many of the common methods, materials, and practices in 3D printing today are not always very sustainable: The printing with metals and thermoplastics requires high energy for the production of a piece, 3D printed concrete causes large CO_2 emissions, bioplastics such as PLA do not biodegrade, and the ubiquity of desktop 3D printing results in excessive waste through the production of test prints and throwaway products.

C. Klemmt (✉)
College of Design, Architecture, Art and Planning, University of Cincinnati, Cincinnati, OH, USA
e-mail: christoph@orproject.com

G. Beaucage
Chemical and Materials Engineering, University of Cincinnati, Cincinnati, OH, USA

© The Author(s), under exclusive license to Springer Nature Singapore Pte Ltd. 2026
C. Klemmt and G. Beaucage (eds.), *Biodegradable 3D-Printing*, The Materials Research Society Series, https://doi.org/10.1007/978-981-95-4025-9_12

The development and implementation of more ecologic materials and methods is therefore paramount in the utilization of the advantages of 3D printing in a sustainable manner.

The intention of this book was therefore the dual aim of bringing together current academic research into additive manufacturing with biodegradable materials, while also promoting more sustainable practices and materials in 3D printing. Researchers from different disciplines have investigated 3D printing with plant fiber, living organisms and other biodegradable materials, and the work is applied across various fields ranging from small scale medical implants to large architectural applications.

While most authors' research is in the development of a specific material or method and shows the results in specific applications, we nevertheless divided the book into those chapters that concentrate on the methods and materials, and those that concentrate on their applications. This allowed us to take a closer look at the possibilities, constraints, and obstacles that various materials and fabrication technologies offer and face, while also allowing us to understand the breadth of possibilities and limitations of their applications.

Lastly, the Appendix offers hands-on information to start using or researching the materials and methods presented in this book, aimed at scholars, students, and enthusiasts alike. It contains detailed information, including possible equipment, recipes, and preparation methods. This is aimed at a broader dissemination and implementation in academia, industry, and crafts.

2 Review of Chapters

2.1 Part 1: Introduction

Chapter 1: Intention of This Book

(Christoph Klemmt, Gregory Beaucage)

This chapter introduced the book and its academic and social intentions (Chap. 1).

Chapter 2: Sustainability in Additive Manufacturing: A Comparison of 3D Printing Methods and Materials

(Rajat Sodhi, Shuo Qin)

The chapter makes a strong case for the development and implementation of more sustainable materials and methods in additive manufacturing. Various data from the Life Cycle Assessments of different materials and additive manufacturing techniques indicate the possibilities as well as challenges of 3D-printing. While waste material is reduced in 3D printing as only the material is printed that is actually required for an object, many technologies have high carbon footprints, or problems of waste at the end of an object's life cycle. High carbon footprints are due to the manufacturing of the raw materials, such as in cementitious materials, as well as due to energy-intensive production methods where thermoplastics or metals must be molten. At the end of an object's life cycle, few materials offer biodegradability, and recycling or industrial composting are often not carried out (Chap. 2).

Chapter 3: Review of Biodegradable Materials in Additive Manufacturing

(Gregory Beaucage, Christoph Klemmt)

Many biodegradable materials have been used in 3D printing and Chapter 1.3 gives a comparative description of a number of these materials particularly those of current interest for research and development. Care is taken to note where claims of biodegradability do not match up with the reality in the US. The chapter describes the use of wood-based materials in some detail. An overview of animate materials is also given (Chap. 3).

2.2 Part 2: Materials and Methods

Chapter 4: FLAM: 3D Printing with Chitin—Cellulose Composites

(Stylianos Dritsas and Javier G. Fernandez)

This chapter presents a method for the 3D printing of cellulose and chitin. Inspired by the formation of objects in nature, the aim was the development of a method that sources widely available materials locally, that uses minimal energy in the production, and that the materials could be naturally reclaimed at the end of their life cycle. The method, based on the two most available biopolymers, was used to construct various objects, including an airfoil and an architectural prototype. The paper points out differences to commonly used materials, including challenges like susceptibility to water or the ability to thermoform. The materials are therefore not suitable for all applications but will help to reduce the reliance on petroleum-derived polymers (Chap. 4).

Chapter 5: 3D Bioprinting with Cellulose-Derived Bioink and Precursors

(Bashir Khoda and Md. Ahasan Habib)

The chapter presents research into extrusion-based 3D printing with living cells, cellulose and its derivatives for medical implants. Focused around the three aspects of design for continuous deposition, bioink and precursor materials, and printing with cellulose and its derivatives, the chapter explains methods to construct larger-scale structures, and the integration of bioinks with sustained viability. While the materials have previously found to be promising for bioprinting due to their low toxicity, the chapter points out the importance of the relationship between design, material and manufacturing in achieving successful results (Chap. 5).

Chapter 6: Living Composites: Mycelium Biofabrication and the Evolution of Design

(David Andréen)

The chapter focuses on 3D printing with mycelium for large-scale applications. The mycelium acts as the binder of a substrate, often composed of lignocellulosic materials. The 3D printing process as well as the curing must ensure the viability of the spores to survive and grow, requiring a close control of the environmental conditions. The author proposes a shift from the "construction" of objects toward their "cultivation," whereby the use of bioprinting of the binding agent results in a novel approach to cradle-to-cradle manufacturing (Chap. 6).

Chapter 7: Three-Dimensional Printing of Lignin and Wood

(Md Shajedul Hoque Thakur and Muhammad M. Rahman)

Another biopolymer that is emerging as a possible biodegradable binder is lignin, which together with cellulose are the main components of wood. The chapter explains processing and utilization of lignin in a variety of 3D printing methods. In some of the methods, the product remains biodegradable, while in others, for example if the lignin undergoes photopolymerization, it does not. The authors then present a method for combining lignin with cellulose, which results in a final product close to 3D printed wood (Chap. 7).

2.3 Part 3: Applications

Chapter 8: Three-Dimensional Printing of Biodegradable Materials in the Biomedical Sector

(Karthik Tappa).

After an introduction of the relevant materials and manufacturing methods, the chapter gives an overview of the applications of 3D printing of biodegradable materials in the biomedical sector. This includes applications in orthopedics, in drug delivery, and in tissue engineering. The materials' behavior of biodegradation is used especially to ensure that implants can disappear without having to be removed surgically, while a continuous degradation can be utilized for a localized and continuous drug delivery. The 3D printing of biodegradable materials can therefore be harnessed for efficient, personalized, and sustainable healthcare (Chap. 8)

Chapter 9: Wet-Mold Concrete Casting with Biodegradable Materials

(Brendan Kiser, Ryan Kennedy, Lucas Palicki)

This chapter presents a method for producing molds for the casting of concrete from biodegradable materials. However, unlike conventional methods, the material is not cured, but remains in its malleable state, with an outer container resisting the hydrostatic pressure, and the concrete being cast simultaneously to the mold extrusion. The mold material can be directly reused or let to biodegrade. By 3D printing molds rather than the final piece, various challenges are overcome, such as the stability of the piece during printing or the integration of reinforcement (Chap. 9).

Chapter 10: Large-Scale 3D Printing of Wood for Architectural Applications

(Christoph Klemmt, Gregory Beaucage)

A method for the large-scale 3D printing of a high wood-content material is presented in this chapter. The material consists of wood fibers with cellulose-derived binders. Unlike most methods, the material has a high wood content above 60% and can be extruded at large bead diameters. The material expands upon its release from the nozzle to counteract shrinkage. A prototype in form of an architectural column has been produced (Chap. 10).

Chapter 11: 3D Printing with Living Matter

(Frank Melendez and Nancy Diniz)

The chapter presents a series of products, including vessels and tiles, that have been 3D printed using living cells, including mycelium and algae. The cells continue to grow after the manufacturing process and become a part of the object. Based on those prototypes, the chapter explores theories of relationships between humans, objects, nature, and the design and production processes. Instead of static objects, the process creates dynamic objects that unfold in time (Chap. 11).

3 Discussion

Although the presented research originates from a wide variety of academic fields and spans various scales, many of the aims, opportunities, and challenges in biodegradable 3D printing have been pointed out by several authors.

3.1 Materials

The type of materials presented in the research, as chosen by the editors of this book, focused on biodegradable materials. Regarding most commercially available materials used in 3D printing, there are significant concerns regarding sustainability. As shown by Sodhi and Qin (2):

> Major concerns exist on the one hand in the energy use for the rawmaterial extraction, processing, and manufacturing, especially using cementitious materials and metals, and on the other hand in the waste management and limited or non-existing circularity, especially in the case of concrete and polymers.

In most cases, the materials can be separated into bulk materials, binders, and solvents. The bulk materials were largely based on plant fibers, including unprocessed wood fiber, cellulose, and cellulose derivatives. Other materials included clay, which, while not biodegradable, occurs extensively in nature and is environmentally benign.

Different binders were presented, including several biopolymers. This included plant materials such as cellulose derivatives and lignin. Other binders in the research included chitin, an animal product, and mycelium, a fungus. Some of the presented research also included non-biodegradable materials such as PLA.

In almost all cases, water was used as the solvent.

3.2 Applications

The presented applications of the biodegradable 3D printing methodologies covered various scales and ranged from small-scale medical implants for drug delivery up to the scale of architectural prototypes. The applications covered academic fields such as biomedicine, aeronautical engineering, product design, art and architecture. The printed prototypes were tested as drug releasing medical implants, temporary scaffolding for bone or tissue growth, design objects, airfoils, molds, and architecture. As Klemmt and Beaucage (10) write:

> It is therefore imperative to reconsider the materials and methods that we use to construct buildings, with a focus on renewable resources, low energy methods, and reusable, recyclable or biodegradable materials.

This wide range of possible applications shows the versatility of 3D printing with biodegradable materials.

3.3 Biodegradation

The biodegradation of the material was pursued for different purposes by the authors: One of the main reasons was the aim to create more sustainable products, by ensuring an ecologically compliant waste disposal. Partly, the approach was application driven, whereby the aim was to produce a specific, commonly not biodegradable product with a more sustainable material or process. On the other hand, some scholars researched the possibilities of utilizing specific biodegradable materials for 3D printing without a specific application in mind.

The other main reason to use biodegradable materials, especially in the medical applications, was the possibility of an intentional and controlled degradation, in the field of biomedicine especially of implants in the human body, which then do not have the be surgically removed after their implantation.

In all cases, the products' end of life was of critical importance and the driver for pursuing the specific production methodologies. As Diniz and Melendez (11) write:

> Efforts regarding conceptualization, functionality, temporality, and aesthetics while working with living and biodegradable materials are pointed toward promoting a decarbonized, regenerative, and symbiotic future in which biological rhythms and processes are embraced. Design experimentation and processes shift from a carefully designed and crafted artifact static in space, to a distributed set of dynamic processes unfolding in time.

3.4 Challenges

Across the spectrum of research presented in this book, also the challenges identified often correlated across scales and disciplines. Tappa (8) states:

> Despite the transformative impact, challenges in material development, regulatory frameworks, and ethical considerations must be addressed to realize the full potential of this technology. Ongoing research and advancements in multifunctional materials, AI-driven design, and sustainable manufacturing hold promise for overcoming these hurdles.

The ongoing as well as future work will therefore have opportunities to address those challenges.

3.4.1 Biodegradation

While the aim of all the authors was to achieve biodegradation at the end of the products' life cycle, biodegradation, or the control of degradation, posed a challenge for several researchers. Some of the applications, such as in the biomedical field, required a tightly controlled degradation over time. In parallel to the biodegradation occurred the release of therapeutic compounds or the regrowth of tissue. A too fast or too slow degradation could therefore adversely impact the healing process.

In other cases, where the project's aim of biodegradation was its sustainable disposal, the degradation was intended to occur at the end of the object's life cycle but not at any time prior to that. As the determination of the object's end of life is often its user's decision, to a certain degree this requires "biodegradation on demand." The biodegradation in many cases commences with a hydrolysis or exposure to other environmental conditions. However, most objects should withstand common environmental condition such as their exposure to water without disintegrating, resulting in the concern of durability vs. biodegradation. Andréen (6) states:

> As these prototypes and experiments accumulate, we begin to glimpse a fundamental shift in how architects and communities conceive of building. Rather than inert, carbon-heavy materials, we see ephemeral, circular cycles of growth and decomposition.

Of importance regarding the degradation are primarily the environmental circumstances in which the objects are kept during their lifespan, and for many of the presented materials especially humidity. Objects produced for protected environments, such as products for indoor use, may have a significant lifespan, while many of the presented materials pose challenges especially for outdoor applications or humid environments.

In a personal anecdote, while I was utilizing starches as binders in 3D printing, I received a note from the facilities manager to please remove my specimens from the building: Mice had started to happily devour them. Just as with traditionally manufactured products made from biodegradable materials, such as wood or paper products, it may be necessary to develop treatments or coatings to extend their lifespan prior to degradation. In the spirit of this book, those treatments should nevertheless still allow the objects to eventually biodegrade.

3.4.2 Design for Continuous Deposition

Within the actual fabrication process, many of the challenges identified are similar to those known from other additive manufacturing methods, such as the design of toolpaths, collapse of uncured structures, or shrinkage. Most of the methods presented in this book are extrusion-based. One issue that was addressed across scales was to extrude continuously so as to avoid stopping and restarting the extrusion at different locations along the toolpath. In the extrusion of liquid slurries, this can easily lead to inaccuracies with an extruder's pressure not immediately ceasing at the nozzle, thereby ejecting further material. Many of the presented projects therefore aimed for a continuous extrusion along the toolpath, and in turn for the design of geometries that could be extruded along a single, continuous path.

3.4.3 Collapse

As many fast-curing additive manufacturing processes, such as melt-extrusion or photopolymerization, result in non-biodegradable compounds, several of the examples in this book rely on processes such as the evaporation of solvents, commonly water, or the growth of mycelium. Those are slow curing, which means that the 3D printed structure must retain its integrity in an uncured state for a significant amount of time to avoid its collapse before a full solidification. This puts constraints on unsupported overhangs and inclined wall segments, and the design of the objects to be printed may need to be adjusted accordingly.

3.4.4 Shrinkage and Deformations

In those research projects that saw a significant volume of solvents evaporate during the curing or drying process, the associated shrinkage and deformations of the 3D printed objects was an inevitability that required addressing. This was evident not only in the projects of this book, but in various papers cited in the reviews of literature. Dritsas and Fernandez proposed to address this by adjustments to the toolpaths, based on a machine learning process that analyzed existing prints (4). Klemmt and Beaucage instead suggested to utilize the expansion of printed fibers to reduce the shrinkage (10). Andreen suggested that a secondary scaffold could be used to stabilize prints, but also to utilize the shrinkage to interlock separate segments during the drying process (6).

3.4.5 Bioprinting

In the projects that utilized bioprinting, including the use of mycelium as the binding agent, a significant concern was the viability of the living cells. Unlike most dead objects, living cells require precise environmental conditions, nutrients, and a toxicity-free environment to survive or thrive. While the biodegradable slurries mostly offered the nutrients and toxicity-free environment, the control of the environmental conditions during and after extrusion were of high importance. As Khoda and Habib (5) state:

> However, challenges persist in achieving mechanical integrity, maintaining structural fidelity, and ensuring desired cellular activities, including viability, proliferation, and differentiation, in printed constructs.

Those methods therefore rely on precisely controlled conditions during as well as after the production process.

3.4.6 Applicability and Commercialization

As with most ongoing research, its applicability and commercialization are usually concerns for the future. Some of the chapters present work that appears close to a possible commercialization, while the work of others examines more fundamental or theoretical principles. Most of the presented projects point out future work that will advance the research and overcome current drawbacks. As Thakur et al. (7) write:

> However, further advancements in processing techniques and compatibility optimization are required to fully unlock lignin's potential in additive manufacturing.

Yet also larger concerns have been raised, regarding issues such as the viability of living cells, permanence of objects, or the acceptance of industry and end-users. As Dritsas and Fernandez (4) state:

> Perhaps one of the greatest challenges with the introduction of a new material process is to overcome the inherent biases from otherwise valuable experience developed working within the parameters of existing technologies. Natural biomaterials do not behave like conventional plastics and forcing them to act like such, for instance being thermoformed and resist oxygen and water, is generally detrimental to their ecological characteristics. They do have, however, unique properties worth exploring for how to make best use of.

A widespread commercial adoption appears to depend to a large extent on the possibilities of the biodegradation: In the cases of a desired degradation during the lifespan of an object, such as in the medical applications, adoptions are likely if the degradation can be controlled well. In applications where objects can be contained in controlled environments to prevent unwanted degradation, adoptions are likely.

However, for applications in exterior, humid, or otherwise degradation-inducing environments, significant challenges are yet to be overcome to facilitate a widespread commercial adoption of the materials.

Beyond the behavior of degradation, most common concerns in material selection apply. While the resources of most plant-based materials are widely available, processing and manufacturing can cause economic concerns. As we pointed out in chap. 3:

> Choice of biodegradable inks depends largely on the cost, intended processing and end-use. For many applications, PHAs are an ideal material but the cost excludes their use from many applications. Wood and wood composites could be useful for large-scale applications but cannot, so far, generate sufficient structural precision for many applications without further working of the printed part.

Nevertheless, all researchers appear very confident that those challenges will be overcome, and that the 3D printing of biodegradable materials has a definite potential in the industry. As Kennedy et al. (9) write:

> With continued development, it has the potential to reshape how architectural components are conceived, manufactured, and realized.

4 Future Developments

The work presented in this book suggests a variety of directions that this specific yet also broad area of research is expected to develop in. Firstly, there is an urgent need to address several of the world's current attitudes toward materials, consumption, renewability, and waste. Mankind has been aware of the thermoplastic industry's waste problem for decades, yet the materials' versatility and the industry's foothold and practices are hard to beat. Nevertheless, it is imperative to explore new avenues and possibilities in manufacturing and material science.

One important aspect is the development of new methods and materials. However, an even more important aspect that the presented, ongoing, and future research need to address are the improvements of the so far developed materials and processes. This covers the full spectrum ranging from sourcing and material preparation to formulations, computational tools in design and processing, production methods, curing and post processing treatments.

Another important aspect will be the development of new applications. While some applications of thermoplastics or other currently used unsustainable materials are difficult to replace, there will also be many applications where their supposed benefits may be of limited use, such as in products with a short lifespan including packaging or giveaways, or where the controlled degradation of a product allows for previously impossible benefits such as a continuous release of drugs in the body.

Future research will also need to include another set of considerations, including financing and supply chains, legal frameworks, ethics, and addressing acceptance by the industry and end-users. The people who drive change like this are often not the academic scientists, but the entrepreneurs who overcome critical stumbling blocks to bring inventions to their commercialization.

We would therefore like to conclude this book with an encouragement for further research and exploration, be it by academics, the industry, students, or enthusiasts. In the appendix, several of the authors have kindly provided detailed formulations and other information that will be helpful in reproducing the presented methods and in taking them further. In a world where the consumption of objects increases ever more, their eventual degradation will be the key to a circular life cycle.

Christoph Klemmt graduated from the Architectural Association in London and received his Doctor Technicae from the University of Applied Arts Vienna. He has worked amongst others for Zaha Hadid Architects, where his responsibilities were focused on the company's projects in China. He is Associate Professor at the Department of Design, Architecture and Planning at the University of Cincinnati, where he set up and runs the Architectural Robotics Lab. He co-founded Orproject, an architect's office specializing in advanced geometries with an ecologic agenda. Orproject has been featured widely and has exhibited at the Palais De Tokyo in Paris, the China National Museum in Beijing and the Biennale in Venice.

Gregory Beaucage is Professor of Chemical and Materials Engineering at the University of Cincinnati. He has published extensively in polymers and materials science with 189 publications and 6,900 citations. He has collaborated with Christoph Klemmt on the development of new materials for 3D printing of wood for the past four years. Beaucage holds a Ph.D. from the University of Massachusetts at Amherst in Polymer Science and Engineering and was a post doc at Sandia National Laboratories before joining the faculty at the University of Cincinnati. He was a Fulbright Global Scholar at Sheffield University in the UK and at several Universities in Ethiopia, South Africa, and Lesotho. He has been Visiting Professor at University of the Witwatersrand, Johannesburg, University of Sheffield, University of Cape Town, and ETH in Zürich.

Appendix

1 Recipe for the 3D Printing of Cellulose and Starch

Christoph Klemmt

This recipe is based on easily available ingredients. It can be used with any 3D clay printer. Depending on the pressure of the printer, the weight percentage of the water can be reduced, resulting in reduced shrinkage.

1.1 Equipment

Item	Specification
Clay 3D printer	Any make, recommended nozzle diameter 1 mm–3 mm
Blender	Ideally a high-shear blender, but a standard kitchen blender can also work for small quantities
Filter	Laboratory filter, possibly a cloth
Hotplate and pot	Any make
Paint mixer	Any make
Bucket	Any make

1.2 Ingredients

Item	Source	wt%
Cellulose fibers, possibly 75% softwood, 25% Hardwood	Paper making supply	7
Starch, corn or potato	grocery	21
Water		72

© The Editor(s) (if applicable) and The Author(s), under exclusive license to Springer Nature Singapore Pte Ltd. 2026

C. Klemmt and G. Beaucage (eds.), *Biodegradable 3D-Printing*, The Materials Research Society Series, https://doi.org/10.1007/978-981-95-4025-9

1.3 Method

Phase 1: Cellulose Preparation

The proposed cellulose fiber mix of 75% softwood and 25% hardwood is standard paper pulp. Cellulose usually comes in one of three forms: Dried in sheets, suspended in water, or loose in fluff form.

The dried sheets can be dissolved in water by means of a blender. The resulting suspension will be similar to acquiring already suspended fibers.

The suspended fibers need to be drained with a filter to remove most of the water. However, some moisture will remain in the fibers, which needs to be accounted for in the overall cellulose and water ratios. It is therefore important to measure the remaining water content and deduct it from the overall amount, especially if a less precise method such as drainage through a cloth is applied. It is recommended that the water content of the fibers is less than a quarter of the overall water content.

Fluff cellulose fibers should be soaked in some water to soften. It is recommended to soak the fibers in less than a quarter of the overall water content, or to drain additional water after soaking.

Alternatively, the cellulose fibers can be obtained by soaking and blending standard paper sheets. However, the paper may contain undesired additives.

Phase 2: Starch Preparation

The starch and remaining water are heated under stirring until the starch has gelatinized.

Phase 3: Combination

Upon cooling, the gelatinized starch and the moist cellulose fibers are mixed. This can be done in a blender, or for larger volumes with a paint mixer.

2 Recipe for 3D Printing Wood Shavings and Methylcellulose

Christoph Klemmt

This recipe is for the co-extrusion of wood shavings with methylcellulose. It requires a 3D printing setup that can extrude larger particle sizes. Extruders for concrete will be able to also extrude this wood slurry.

In the presented form, the dried ingredients contain 50%wt wood, and 100%wt plant products. Depending on the pressure of the extrusion mechanism, the wood content can be increased. The binder consists of two similar compounds, but with differing viscosity. Therefore, the viscosity of the slurry can be tuned by adjusting the ratio of methylcellulose to hydroxypropyl methylcellulose.

2.1 Equipment

Item	Specification
Concrete 3D-Printer	
Concrete or paint mixer	Electric handheld, hopper-type grout mixer, or planetary stand mixer
Sealed container	

2.2 Ingredients

Item	Source	wt%
Pine wood shavings, dimensions 1–5 mm × 5–20 mm × 0.01–1 mm	Animal bedding or saw mill	12.5
Methylcellulose (4000cps@2%)	Pure Original Ingredients	6.2
Hydroxypropyl Methylcellulose (15cps@2%)	Bulk supplements.com Hypromellose powder	6.2
Water		75.1

2.3 Method

Phase 1: Soaking of Wood

The wood needs to be soaked in water to soften so that it does not clog the extruder. It should be stored in a sealed container with 67% of the total amount of water for a minimum of 3 days. The extrudability of larger or thicker wood particles will depend on the possibilities of the extruder.

Phase 2: Binder

The viscosity of the binders, given in cps @ 2% solution, is of importance to achieve a printable viscosity of the slurry. The binder is added to the moist wood shavings and mixed in with a handheld concrete mixer. After the slurry is mixed well, the remaining water is added and stirred in. The slurry remains to soak in the sealed container for at least one more day.

3 Recipe: Pulp Faction BioFDM

Ana Goidea and David Andréen

Published under CC BY 4.0 license.

This instruction is for the bioFDM printing process developed at bioDigital matter lab. It was developed for the production of Protomycokion, a 2 m tall experimental column. The recipe uses a wood-decomposing brown-rot fungus, *Gloeophyllum,* and results in lightweight objects with a hard, brownish and variable surface. The resulting surface is hydrophobic yet allows water vapor permeability.

3.1 Prepare Fungus Culture

The fungus is propagated in a liquid culture using malt extract.

3.2 Mix Substrate for Inoculation

The substrate (pre-inoculation pulp) is mixed to a homogenous paste consisting of:

a. Water (60wt%)
b. Wood particles (18wt%). Wood particle size around 2–4 mm optimizes extrusion consistency and structural integrity. Smaller particles may impede growth, while larger particles can cause nozzle blockages. This recipe is tested for pine wood, but other timbers may be equally appropriate.
c. Clay powder (17wt%). Our recipe uses kaolin clay, other clays affect material behavior and color. The addition is important for the printability of the paste.
d. Cellulose fibers (5wt%). Our recipe used long fiber paper pulp; shorter fibers from recycled paper may likely improve printability.

The ingredients should be as clean/sterile as possible before mixing. If the wood particles are desiccated from sterilization, additional water may be needed to compensate.

3.3 Inoculation

The substrate is sterilized and inoculated with live fungus, after which the mixture is left to grow in sterile conditions. Initial sterilization ensures the fungus establishes dominance in the substrate, significantly reducing the risk of contamination during the subsequent non-sterile printing process.

a. Sterilize the substrate via autoclave or equivalent method (such as heating to lower temperature for longer time).
b. Liquid *Gloeophyllum* culture is blended and added to substrate. Typically, add approximately 1% by weight of the prepared fungal liquid culture relative to the substrate mass.
c. Allow the inoculated pulp to grow under sterile and humid conditions (>90% humidity and adequate oxygenation) at 25 °C for about 3 weeks to establish robust fungal colonization.

3.4 Printing

Note: At this stage it is important that the fungus is well established in the pulp, as the printing process is not sterile, and weak establishment leaves the pulp susceptible to contamination. During mixing and printing, it is imperative to minimize exposure to contaminants—keep equipment sterilized, avoid air currents and falling debris.

a. Additional xanthan gum (2–3% by weight) is added to the pulp which is mixed to homogenous consistency and placed in extraction containers. Water content may need to be adjusted at this stage if it is too dry.
b. A clay 3D printer such as the Lutum 5 is suitable for printing the pulp.
c. The print bed should allow for free airflow, while providing a stiff base that doesn't deform from the weight of the print. A honeycomb extrusion fixed to a stiff frame is recommended.
d. The pulp recipe is suitable for an extrusion width of 3–5 mm, and should allow airflow around all extruded surfaces. Thicker geometries will reduce colonization and strength, and are difficult to dry optimally. The wet pulp is stable in print heights of up to 100 mm, depending on other variables (notably printed geometry).
e. After printing, a honeycomb extrusion or mesh is applied to the top of the print to maintain stability during growth phase, and to constrain shrinkage due to drying to the vertical dimension.

3.5 Growth

a. Store printed structures in a humid, oxygenated environment, sealed to minimize contamination. Maintain relative humidity above 90% and ensure adequate filtered ventilation to sustain oxygen levels without drying the pulp.
b. Allow 3–6 weeks of growth. Too little growth leads to weak binding, and too much results in excessive decomposition of the wood.

3.6 Drying/Sterilization

a. The prints are dried in a desiccated and heated environment with fan-driven air circulation. Too rapid drying may result in delamination and distortion.
b. The prints are then treated at high heat (70 °C) to kill the fungus.

Note: It may be possible to achieve secondary bio-welding if (b.) is skipped, and the fungus remains alive but dormant through the drying process.

3.7 Post-processing and Assembly

a. The dried components can be treated by sanding or other methods and assembled into larger superstructures at this stage.
b. At joints, it may be possible to reactivate the fungus locally by wetting it, and/or applying more inoculated pulp at the joints, thereby bio-welding the sub-components to a heterogeneous material. In this case, step ***3.6*** *b.* should be skipped.

4 Rearing *Hermetia Illucens* for the Extraction of Chitin and Production of Chitosan

Javier G. Fernandez

This protocol outlines the complete circular process of rearing Black Soldier Fly (BSF, *Hermetia illucens*), processing them into biomass, extracting chitin, and producing chitosan. The final product is a high-quality chitosan suitable for film formation and FLAM-based biomanufacturing.

4.1 Equipment

Equipment	Notes
Rearing trays with mesh lids	Any
Climate-controlled insect rearing chamber	Any, not necessary in tropical and subtropical regions between latitudes 40°S and 45°N
Food waste processor	Any, depending on the desired out-put, a commercial blender would work
Freezer (−20 °C or lower)	Any
Hydraulic press or centrifuge	5–20 tons is sufficient

(continued)

(continued)

Equipment	Notes
Laboratory oven	Any, only used for drying at 40–60 °C
Three-neck flask with condenser	Any
Magnetic stirrer with hot plate	Any
Buchner funnel + vacuum filtration	Any
Sieve (100 μm)	Any
pH meter	With prove for constant monitoring

4.2 Materials

Material/Ingredient	Source	Use
Organic food waste	Local food industry/kitchens	BSF substrate
BSF eggs or larvae	Commercial supplier or existing colony	Starting colony
Hydrochloric acid (1N)	Commercial supplier	Demineralization
Sodium hydroxide (1 M)	Commercial supplier	Deproteinization/deacetylation
Glacial acetic acid	Commercial supplier	Chitosan dispersion
Sodium acetate	Commercial supplier	Viscosity solvent system
Distilled water	Laboratory supply	Washing steps

4.3 Directions

PART A: Rearing of Black Soldier Fly (BSF)

1. Place BSF eggs or newly hatched larvae in plastic trays with mesh covers and ventilation.
2. Use organic waste (e.g., spent grain, okara) and homogenize to 65–70% moisture.
   **Do not over-wet the substrate. Anaerobic conditions harm larvae.*
3. Maintain temperature at 27–30 °C, RH at 60–70%, 12:12 light/dark cycle.
4. Feed for ~ 12–14 days until prepupal stage.
   **The prepupal stage (i.e., harvesting time) is marked by the larvae darkening and starting to exit the substrate.*
5. Separate larvae from the substrate using a 100 μm sieve and wash.

PART B: Press-Cake Preparation

6. Freeze larvae at -20 °C for at least 12 h.
   **Freezing at -20°C is regarded as a humane and ethical euthanasia method for insects, aligning with ethical standards in scientific research to minimize potential animal suffering. Ensure the method complies with local regulations.*
7. Thaw and grind into a paste using a food processor or commercial blender.
   **Optionally defat using a cold hydraulic press or centrifuge.*
8. Collect the solid fraction (press-cake) and store it at -20 °C if long-term preservation is required.

PART C: Extraction of Chitin

9. Suspend press-cake in 1N HCl and stir for 2 h at room temp for demineralization.
10. Filter through a 100 μm sieve and wash to neutral pH.
11. Treat with 1 M NaOH at 80 °C for 12 h for deproteinization. Repeat two more times.
    **First deproteinization will yield a brown byproduct that will become clear in subsequent repetitions.*
12. Filter, wash to neutral pH, then dry at 40 °C for 12 h. Store in sealed containers to avoid water uptake.

PART D: Deacetylation to Produce Chitosan

13. Treat chitin with NaOH (1:20 w/v) at 120 °C for 8 h in reflux setup.
14. Cool, wash to neutral pH with distilled water, and dry at 40 °C for 12 h.

PART E: Chitosan Film Production (Optional)

15. Dissolve 1–2% chitosan in 1% acetic acid, stir overnight.
16. Pour into molds and dry to form films.

4.4 Notes

- Consistent waste quality and stable climate conditions enhance BSF growth.
- Typical chitin yield is 6–9% of BSF dry mass; deacetylation yields ~ 87% chitosan.
- Chitosan is suitable for biodegradable packaging, 3D printing (FLAM), and biomedical applications.

5 3D Printing Wood Recipe

Muhammad M. Rahman

5.1 Equipment

Item	Specification
Mixer	Thinky planetary centrifugal mixer (model ARE 310 or similar)
DIW 3D printer	Hyrel engine HR or system 30 M or any other DIW printer
Extruder	SDS-60XT EXTRUDER or any extruder of similar torque
Syringe	Luer-lock syringe
Nozzles	Smooth-flow tapered tips (Nordson EFD or similar) of 25 gauge
Slicer software	PrusaSlicer
Printing substrate	Silicone

5.2 Method

Ink Formulation

Organosolv-fractionated hardwood lignin is thermally reprocessed at 180 °C for devolatilization of gaseous products and residual solvent. Specifically, powder lignin is sheared in a 350-cm^3 batch mixer of a Brabender polymer processing system at 180 °C, 60 rpm rotor speed for 60 min to obtain devolatilized chunks. These chunks are then pulverized until they can pass through a 120-mesh sieve. TOCN and 10.6 wt.% aqueous solution of CNC were mixed with lignin at a ratio of 15:142:10 by weight of TOCN, CNC (aqueous), and lignin, respectively. Then, the mixture was placed in a THINKY Planetary Centrifugal Mixer with two 8 mm steel bearing balls, mixed in three cycles at 2000 rpm for 5 min, and then defoamed at 2200 rpm for 1 min. Last, the mixture is taken out and is ready for printing.

3D Printing

The ink is then loaded in a Luer-lock syringe and shaken to ensure bubbles are removed, and then the nozzle is attached. The ink is printed on a silicone substrate to prevent the adhesion of the wood structure and build plate and facilitate detaching the printed object from the build plate. The ink flow rate is tuned to facilitate smooth printing.

Freeze Drying

The printed samples were placed under dry ice for 30 min and then freeze-dried at −85 °C, under less than 0.1 mbar pressure, for 48 h using the FreeZone 4.5 Liter Benchtop Freeze Dry System.

Heat Treatment

The freeze-dried samples were heated at 180 °C to get the final 3D printed wood part.

6 Creation of G-code for 3D Printing with Open Source Platform

Zafir Bashir andBashir Khoda

Project Link: https://scratch.mit.edu/projects/1161896381/editor

Commonly available 3D printing software (e.g., CURA) works well for their predefined and preprogrammed infill patterns. However, for successful bioprinting, biotypology optimized infill pattern may be necessary to achieve the functionality as discussed in the chapter. In this section, the methodology to create variational and controllable 3D infill patterns will be demonstrated using the common open-source platform Scratch (scratch.mit.edu).

G-code is a low-level machine language that utilizes a set of instructions for the computer numerical control (CNC) system which are commonly used in extrusion based 3D printers. The G-code instructions start with 'G' followed by a numerical value, for which each number represents a specific action. For example, G0 is for Rapid Transverse; G1 is for Linear interpolation; G2 and G3 are for Circular interpolation, CW and CCW, respectively. For 3D printers, G1 is the most common instruction that moves the motors resulting in 3-axis actuation. Code E sends the instruction for the extruder motor that purges a desired material volume through the nozzle. An example of a common block is:

G1 X2 Y3 Z4 E0.50:: *Move to destination coordinate (2,3,4) while depositing 0.5 unit volume of materials.*

Following such a single line block in sequence, the printer moves and deposits material in layers to print the object. Few other auxiliary instructions are also used for 3D printing such as F for nozzle speed; M-codes are for setting up auxiliary parts of the 3D printer, not used for the movement of the extruder. Depending upon the

Z. Bashir
Bangor High School, Bangor, ME, USA

B. Khoda
Mechanical Engineering Department, University of Maine, Orono, ME, USA

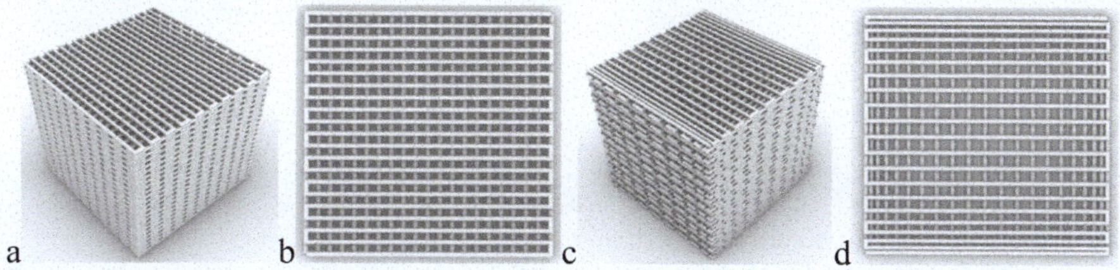

Fig. 1 **a–b** Uniform infill isometric view, top view and **c–d** non-uniform infill isometric view, top view

Fig. 2 Zigzag infill pattern with **a** Horizontal, **b** Vertical lines

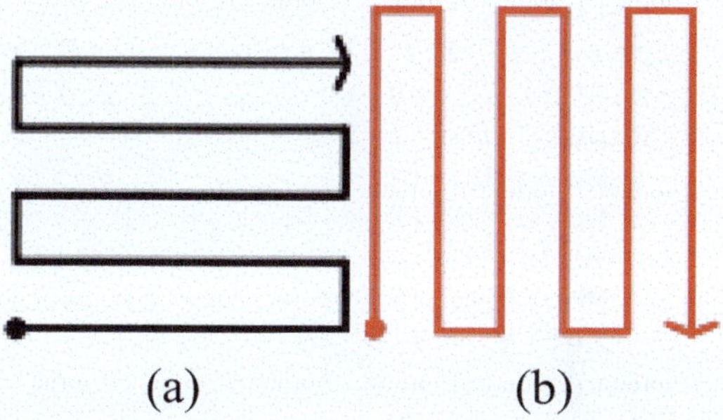

size and shape of the object, hundreds of lines or blocks are required for printing the object. This guide will demonstrate how to generate these repeatable blocks by utilizing Scratch, a free and basic coding platform, to print desired infill patterns on the Ender 3 S1 Pro printer. For demonstration purposes, we will explain the steps to generate two 3D printing G-code with (i) uniform and (ii) variational infill as shown in Fig. 1. A zig-zag infill pattern (Fig. 2) will be considered for this demonstration.

6.1 G-Code Generating Algorithm

Input

- Point Location: P(X, Y, Z)
- Extrusion Thickness: T // depends on nozzle diameter and print speed
- Filling Space Dimensions: L (Length), W (Width), H (Height) // number of layers is H/T
- Infill Spacing: S // S can be constant or a function S = f(α, ß, …) which is a function of α,ß.. as discussed in the chapter.

Output

See Fig. 2.

```
Move to (X, Y, Z)  // Go to the starting point
Extrude, E=(Pi-Pi-1)*vf     // volume factor 'vf' is a
constant related to the desired quality
Repeat (H/T):
  If (Z / T) is even:  // Zigzag Infill Pattern with
Vertical Lines (c.f., Fig 2a)
    Repeat for i from 0 to (L / S) - 1:
      Move to (X, Y + W, Z) while extruding E
      Move to (X + S, Y + W, Z) while extruding E
      Move to (X + S, Y, Z) while extruding E
      Move to (X + 2S, Y, Z) while extruding E
      Set X = X + 2S
    Return to the Starting Point and Set Z==Z+T
  Else: // (Z / T) is odd, \\ Zigzag Infill Pattern with
Horizontal Lines (c.f., Fig 2b)
    Repeat for i from 0 to (W / S) - 1:
      Move to (X + L, Y, Z) while extruding
      Move to (X + L, Y + S, Z) while extruding
      Move to (X, Y + S, Z) while extruding
      Move to (X, Y + 2S, Z) while extruding
      Set Y = Y + 2S
    Return to the Starting Point and Set Z==Z+T
```

6.2 Implementing with Scratch

Scratch is a beginner-friendly, block-based programming language developed by the MIT Media Lab. It is block-based and allows the user to create programs easily without the hassle of memorizing functions or syntax. The functions are categorized by color, representing key aspects like the variables, control, and operators. To implement the algorithm above, we will be utilizing the motion blocks (blue) (Fig. 3), the functions (light red), the lists (dark red), the variables (orange), and the operators (green). We will also need the "pen" blocks, which can be found by clicking the extension button at the bottom left and selecting "pen" (Fig. 4).

The pen tool creates a visualization of the generated 3D printable structure. Although this is optional and does not affect the output of the G-code, it is a good way to troubleshoot errors in the code.

6.3 Introducing the Variables

Scratch has built-in sliders under the "variables" tab as shown below (Fig. 5), which can be used to provide the input values defined in the algorithm. It should be noted

Fig. 3 Motion overview

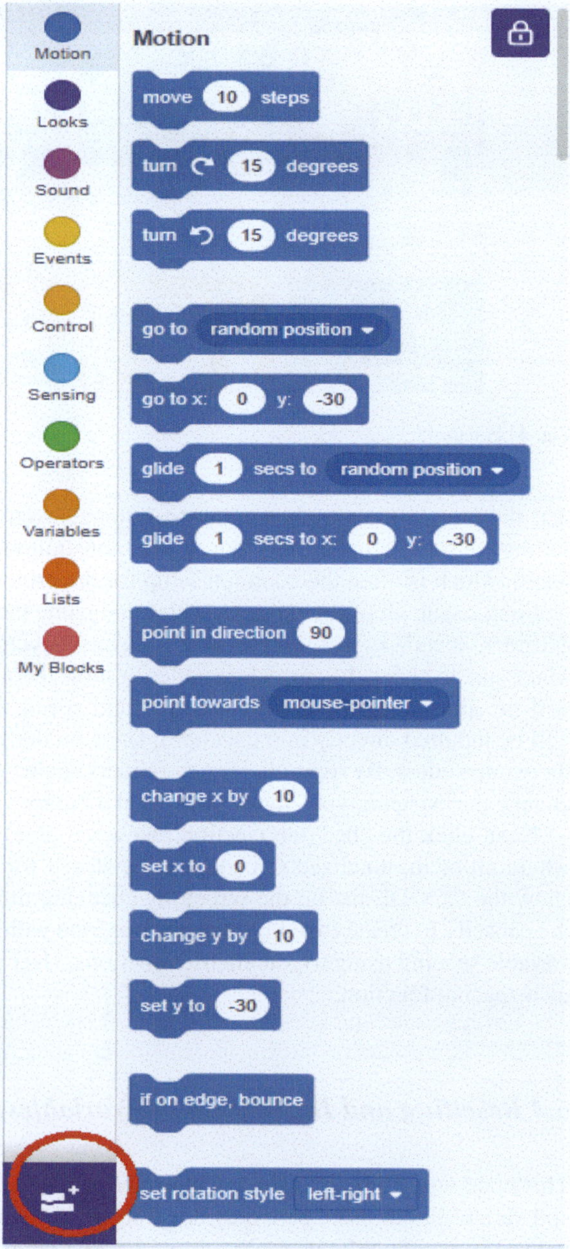

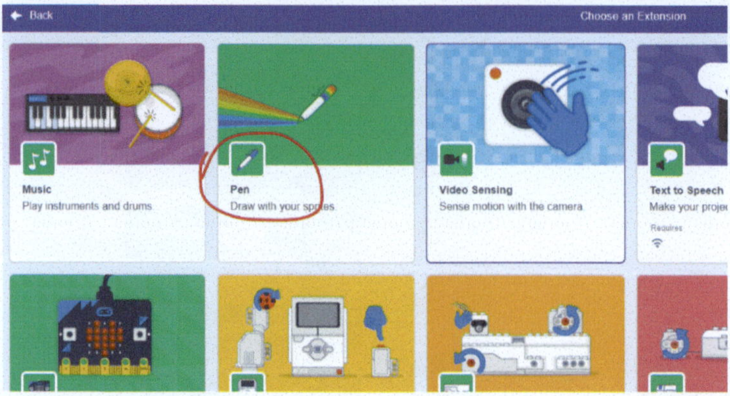

Fig. 4 Pen tool

that the unit of the variables must be homogenized and matched with the printer settings (e.g., FPS or SI unit). For this demonstration, "millimeter" will be used for length which is often the common setup for the printer.

First, create all of the input variables and name them accordingly (e.g., distance, extruder, speed, x, y, z, length, width, layers, layercount, and spacing variables). Make sure to check the correct variables to show them on the screen. These variables will be adjustable, along with the x, y, and speed variables (the values of which will be inputted directly into the code). Once on the screen, right-click them to turn them into sliders. By right-clicking the sliders again, you can change the range. Only change the "spacingtype" range, making it a binary variable from 0 to 1.

Next, click the "list" tab (section below variable) and name it GCODE. This is where all of the finalized G-code will be stored for downloading. Make sure you show the GCODE list on the screen by checking the box next to it. This code has the capacity to create two different patterns, one with uniform spacing and one with variable spacing using trigonometric functions. User will also be able to define their own spacing function.

6.4 Resetting and Initializing the Variables

This code will reset all of the uncontrolled variables. You can change the x, y, and spd variables, but make sure they are compatible with your 3D printer (no negative values, correct coordinate range, etc.). The z variable is held constant at 0.2 because that is the approximate height of the filament for a common 0.4 mm nozzle. The "spd" variable is the relative time it takes the visual simulation to run. The period variable is held constant at 1.25 * spacing. This ensures a symmetric structure, but it can be changed for asymmetry.

Fig. 5 Variables tab

The "delete" block removes any previous code in the list. The "add" block creates the first line in the code, using the G0 function, which is a rapid move command often used for non-deposition travel. Extruding volume code E is an accumulating quantity and G92 is used to reset the extruder value. Joining that with "F1200" sets the initial speed of the nozzle at 1200 mm/min heading towards the initial predetermined coordinates (0, 0). At the very end of the script are the pen and motion blocks, which are solely for the visualization (Fig. 6).

6.5 Calculating the E Value with Function Block

Now create a function block (the "My Blocks" category below the lists) to calculate the extrusion value, calling it "E Calc". Make sure to tick "run without screen refresh", ensuring that the function runs instantaneously with minimal delay. Add an input and call it x/y. This function will calculate the distance traveled along the x and y axes and then calculate the extruder value by multiplying the distance by 1/30, a volumetric factor for the 0.4 mm nozzle. This value may differ depending on filament size, density, and how thick/thin the print should be. Because the extruder value is

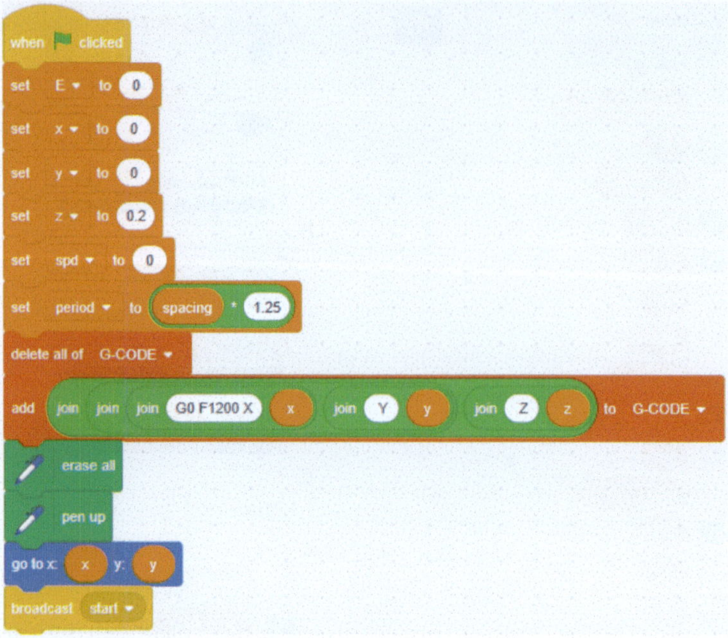

Fig. 6 Resetting and initializing the variables

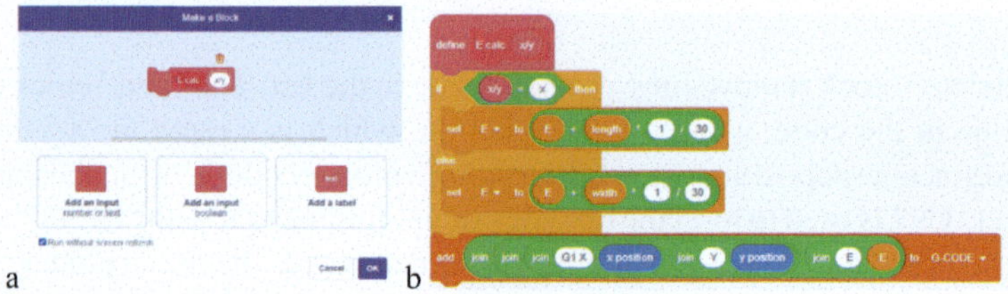

Fig. 7 **a** Block input. **b** Flow diagram

compounding, it will add it to the previous value. These functions will be called when constructing each layer of the lattice structure (Fig. 7).

6.6 Example Code

To demonstrate, Fig. 8 shows the code for the G-code generation of one line.

After the initial code, the program moves to a set of coordinates determined by the input variables and converts its position into G-code, creating a line. The top two blocks are for the visualization while the bottom two blocks are for the actual

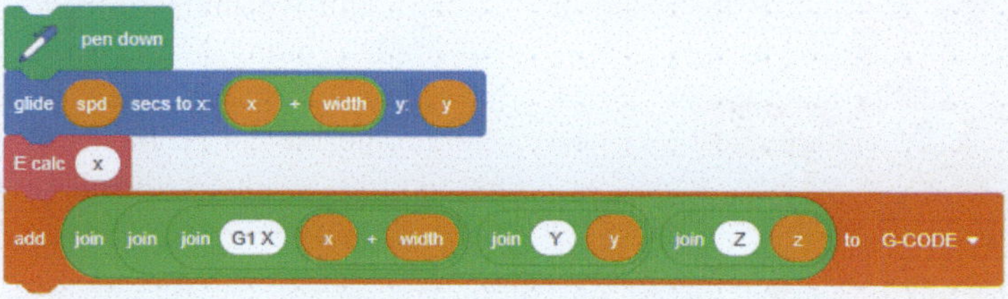

Fig. 8 G-code generation for one line

G-code generation. Since the structure we are trying to create is compiled out of various lines, this code is the foundation of the program.

6.7 Calculating the Spacing

Next, create another function called "spacing calc", which will calculate the necessary spacing value depending on the initial input and whether it is a uniform or variable spacing type. When spacingtype is set to zero, the program will output a uniformly spaced structure, and vice versa. For the variable structure, this program will output one based on the cosine function, though this can be changed simply by changing the cosine dropdown to a sine dropdown. Also, create a short function named "reset var" which will help to calculate the distance traveled. Tick the "run without screen refresh" box as well (Fig. 9).

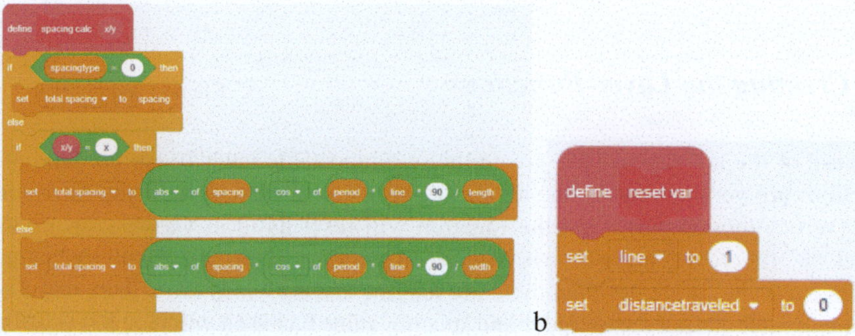

Fig. 9 Calculating the spacing

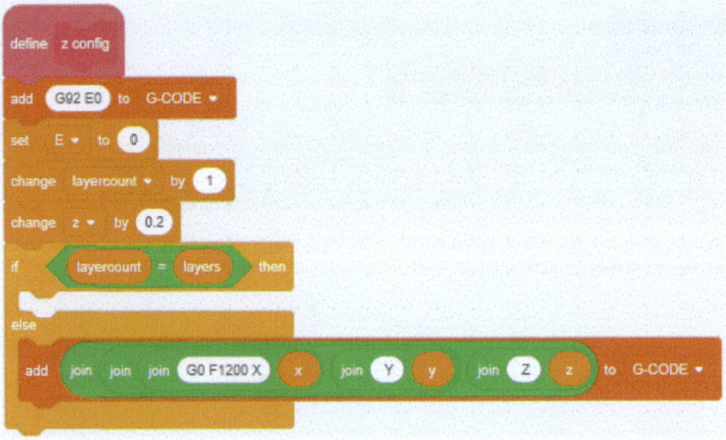

Fig. 10 Changing layers

6.8 Changing Layers

We still need to increase the z-axis by the extrusion thickness (T) after every layer. Not doing so would result in the printer continuously printing on the same plane, causing the layers to merge and potentially cause damage rather than forming a three-dimensional structure. To raise the z-axis, create another function and check the "run without screen refresh" option. After every layer, this function will be called. It will change the layer count by one and keep increasing the z-axis until the program reaches the desired number of layers (Fig. 10).

6.9 Creating the Layer Functions

Because of the zig-zag pattern, create the two repeatable layer functions. The two functions are very similar to one another, with the only difference being the direction (vertical/horizontal). First the function will reset its local variables using the "reset var" function. Then a loop is introduced, which will repeat until the calculated distance traveled is greater than the input length. The code will draw lines based on the inputs, calculating the extruder and spacing value based on whether it is uniform or variable (Fig. 11).

Figure 12 is the second function, which creates horizontal layers rather than vertical

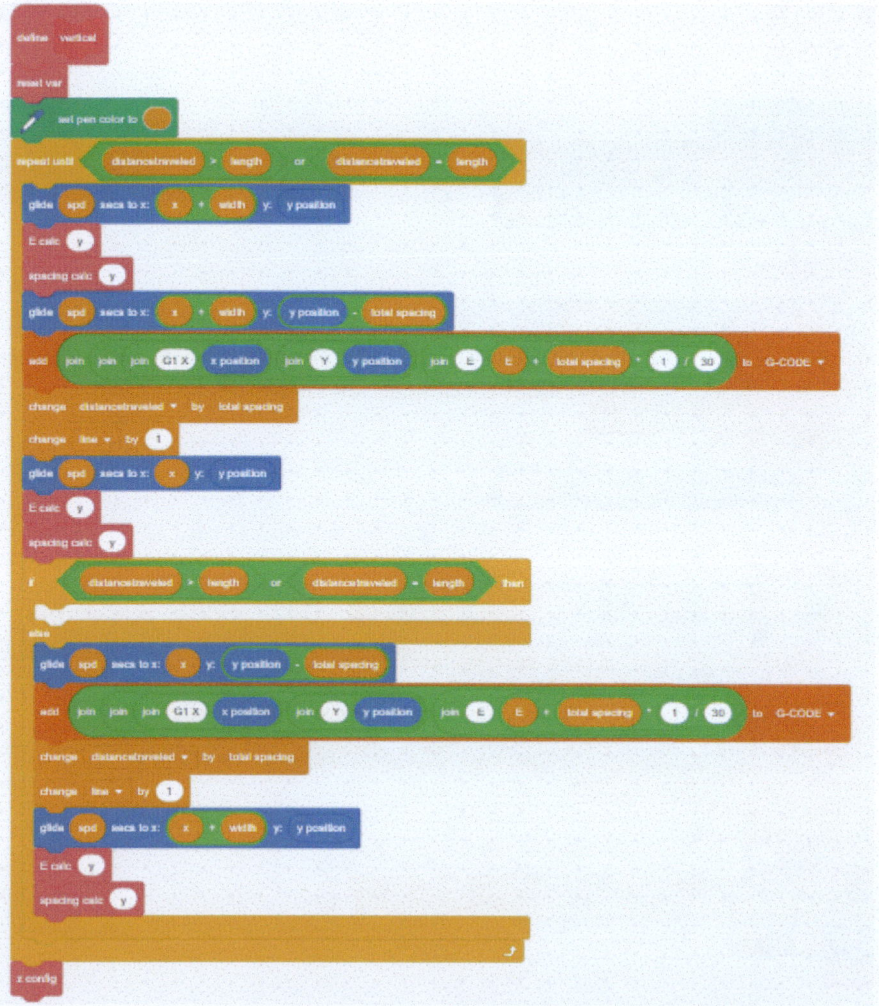

Fig. 11 Creating the layer functions

6.10 Combining and Calling the Functions

Finally, it is just a process of calling each function properly and in order. The "when I receive start" block is an indication of when the variables are set, and the program can run. This is usually done by creating a button that broadcasts this message after being clicked. Alternatively, you can add this code under the setup code. The rest of the code determines the number of times each layer function should be run, based on the number of layers using the modulus function. For example, if I wanted five layers, this code sets the number of times both layers should be run (twice; 5 mod 2) and how many extra layer functions need to be called (based on the remainder of

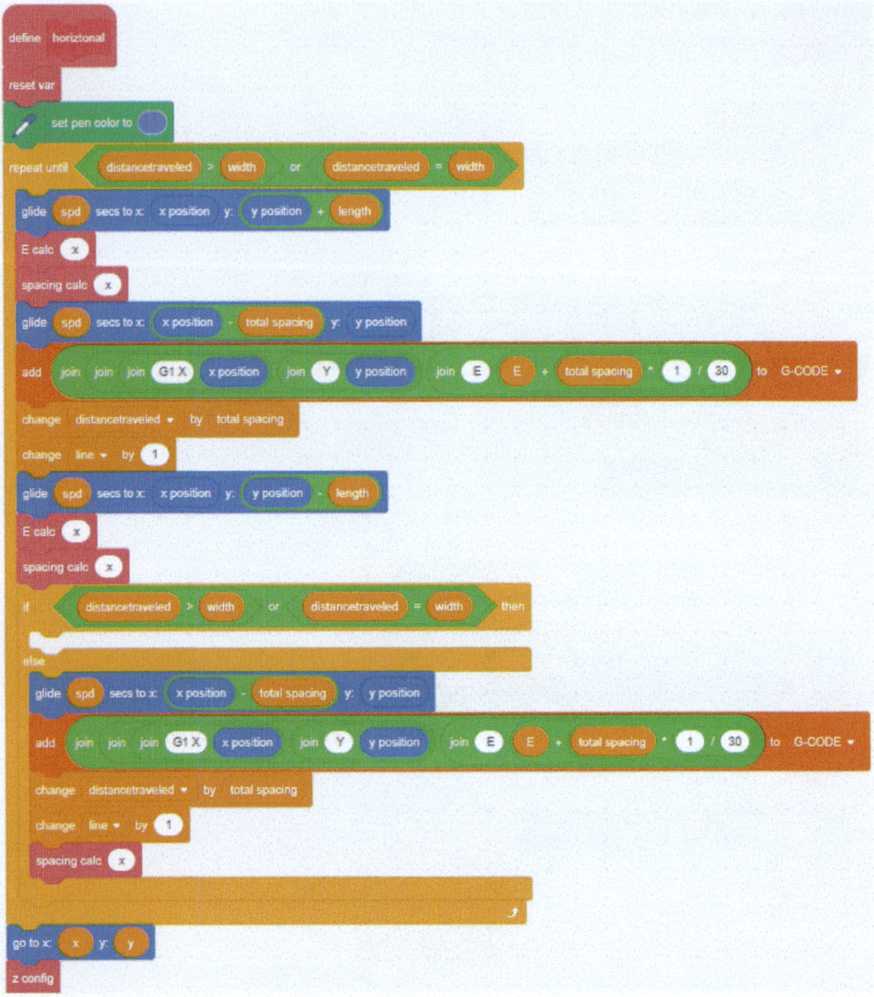

Fig. 12 Horizontal layers rather than vertical

one). After this, the program should run and successfully generate the main section of G-Code based on the set of customizable variables (Fig. 13).

The code can be saved by right-clicking on the list and clicking "export". It should be noted that a printed specific header and footer code must be added to our generated code so that it can control the 3D printer of our choice (Fig. 14).

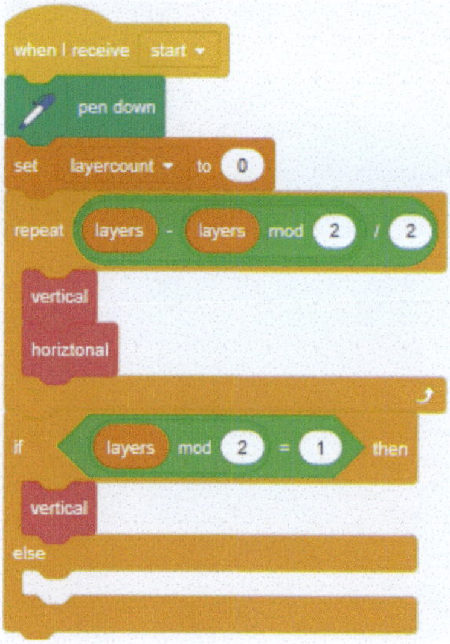

Fig. 13 Combining and calling the functions

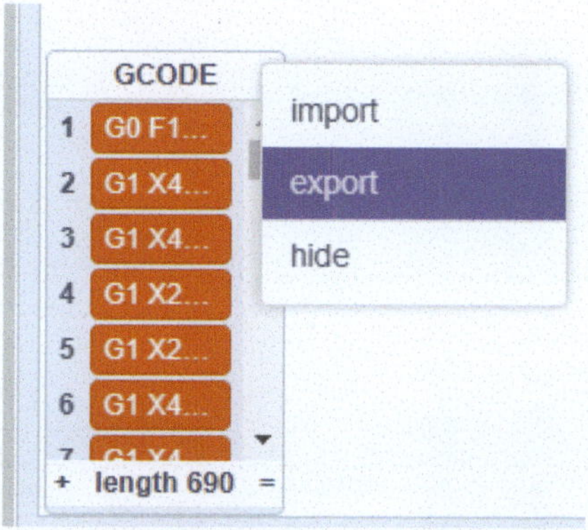

Fig. 14 Export of the code

Index

© The Editor(s) (if applicable) and The Author(s), under exclusive license to Springer Nature Singapore Pte Ltd. 2026

C. Klemmt and G. Beaucage (eds.), *Biodegradable 3D-Printing*, The Materials Research Society Series, https://doi.org/10.1007/978-981-95-4025-9

O

P

R

The manufacturer's authorised representative in the EU is Springer Nature Customer Service Centre GmbH, Europaplatz 3, 69115 Heidelberg, Germany. If you have any concerns regarding our products, please contact ProductSafety@springernature.com

Printed and bound by CPI Group (UK) Ltd, Croydon, CR0 4YY
07/10/2026
02229350-0002